AF389396

Plant Polyamines

Plant Polyamines

Special Issue Editor
Taku Takahashi

MDPI • Basel • Beijing • Wuhan • Barcelona • Belgrade • Manchester • Tokyo • Cluj • Tianjin

Special Issue Editor
Taku Takahashi
Okayama University
Japan

Editorial Office
MDPI
St. Alban-Anlage 66
4052 Basel, Switzerland

This is a reprint of articles from the Special Issue published online in the open access journal *Plants* (ISSN 2223-7747) (available at: https://www.mdpi.com/journal/plants/special_issues/plant_polyamines).

For citation purposes, cite each article independently as indicated on the article page online and as indicated below:

LastName, A.A.; LastName, B.B.; LastName, C.C. Article Title. *Journal Name* **Year**, *Article Number*, Page Range.

ISBN 978-3-03936-318-6 (Pbk)
ISBN 978-3-03936-319-3 (PDF)

Contents

About the Special Issue Editor

Taku Takahashi, Ph.D. Taku Takahashi has been serving as a Full Professor of the Division of Earth, Life, and Molecular Sciences, Graduate School of Natural Science and Technology, Okayama University, Japan, since 2004. He completed his Ph.D. at the University of Tokyo in 1992, after which he worked as a postdoctoral researcher at the Rockefeller University (1992–1995) and as an Assistant Professor at Hokkaido University (1995–2004).

Editorial

Plant Polyamines

Taku Takahashi

Graduate School of Natural Science and Technology, Okayama University, Okayama 700-8530, Japan; perfect@cc.okayama-u.ac.jp; Tel.: +81-86-251-7858

Received: 8 April 2020; Accepted: 13 April 2020; Published: 16 April 2020

Abstract: Polyamines are small organic compounds found in all living organisms. According to the high degree of positive charge at physiological pH, they interact with negatively charged macromolecules, such as DNA, RNA, and proteins, and modulate their activities. In plants, polyamines, some of which are presented as a conjugated form with cinnamic acids and proteins, are involved in a variety of physiological processes. In recent years, the study of plant polyamines, such as their biosynthetic and catabolic pathways and the roles they play in cellular processes, has flourished, becoming an exciting field of research. There is accumulating evidence that polyamine oxidation, the main catabolic pathway of polyamines, may have a potential role as a source of hydrogen peroxide. The papers in this Special Issue highlight new discoveries and research in the field of plant polyamine biology. The information will help to stimulate further research and make readers aware of the link between their own work and topics related to polyamines.

Keywords: copper amine oxidase; polyamine oxidase; spermidine; spermine; stress response; thermospermine

---

Polyamines are polycationic compounds ubiquitously found in all organisms and play various roles by interacting with negatively charged macromolecules, such as DNA, RNA, and proteins. They are also present as a conjugated form with organic molecules and proteins. In plants, polyamines have been shown to regulate growth, development, and responses to biotic and abiotic stresses. In recent years, our understanding of plant polyamines in terms of their biosynthesis, metabolism, and physiological function has seen major advances [1–3]. In this situation, I decided to edit a Special Issue aimed at highlighting new discoveries and advances in plant polyamine research.

One unique feature of plant polyamines is the widespread occurrence of thermospermine from primitive algae to land plants [4]. However, it was only after the turn of the century that the specific function of thermospermine in regulating xylem development of vascular plants was discovered. The gene for thermospermine synthase had initially been identified as the gene for spermine synthase from a mutant of Arabidopsis, which shows an excess xylem development with a dwarf growth phenotype, but it was later proven to code for thermospermine synthase [5]. In accordance with its vascular-specific expression, *AtPAO5*, a gene encoding polyamine oxidase (PAO) that predominantly catalyzes thermospermine, also shows vascular-specific expression in Arabidopsis [6–8]. Sagor et al. [9] report on their finding that the level of thermospermine and the hypersensitivity to thermospermine in the Arabidopsis *pao5* mutant is restored by transgenic expression of its homologue *SelPAO5* from *Selaginella lepidophylla*, a primitive vascular plant related to ferns. Interestingly, SelPAO5 converts thermospermine to norspermidine instead of spermidine. On the other hand, our study focused on the effect of exogenous thermospermine on vasculature in a monocot plant and found that it has a repressive effect on xylem development in rice crown roots [10]. The study also revealed that the genes encoding apoplastic PAOs are drastically activated by both spermine and thermospermine.

Unlike other plant growth regulators, which generally control the expression of genes via activation of specific transcription factors, polyamines have been shown to act in mRNA translation. A paper by

Poidevin et al. [11] clearly and concisely reviews the involvement of spermidine and thermospermine in regulating mRNAs at the level of translation. The authors discuss the significance of these polyamines in the context of mRNA quality control.

There is accumulating evidence that polyamine oxidation, the main catabolic pathway of polyamines, may have a potential role as a source of hydrogen peroxide (H_2O_2). A review by Yu et al. [12] provides recent information on plant PAOs, especially those in three model species: rice, Arabidopsis, and tomato. Toumi et al. [13] addressed a potential link between polyamine catabolism and heat-shock proteins, HSP90s, under heat stress using Arabidopsis mutants in combination with pharmacological experiments and revealed that heat stress-induced increase in acetylated spermidine and spermine is further enhanced in plants having reduced levels of HSP90s. Because the increased polyamine titers under heat stress were associated with the increase in PAO-derived H_2O_2, the authors suggest an antagonistic relationship between HSP90s and polyamines/PAOs/H_2O_2 homeostasis.

The Arabidopsis genome contains two genes for spermidine synthase, one gene for spermine synthase, one gene for thermospermine synthase, and five genes for PAOs, that exhibit different or overlapping substrate specificity [10]. By contrast, there are ten putative genes encoding copper amine oxidase (CuAO) annotated in the Arabidopsis genome [14]. This might reflect the functional divergence of this enzyme. CuAOs catalyze the intracellular and extracellular terminal catabolism of amines, including monoamines, diamines and polyamines [15]. The expression of these genes is inducible by stress-related hormones such as methyl-jasmonate, abscisic acid (ABA), and salicylic acid, or by wounding, and they have been shown to function in wound healing, defense against pathogens, protoxylem differentiation, and stomatal closure [16]. However, not all genes have been characterized in terms of the physiological roles so far. A study by Fraudentali et al. [17] addresses the function of AtCuAOδ and provides data showing that AtCuAOδ is involved in the H_2O_2 production related to ABA-induced stomatal closure. Recently, the authors also revealed that leaf-wounding induces expression of another CuAO gene, *AtCuAOβ*, in root vascular tissues [18].

Polyamines are present in cells at relatively high concentrations in the mM range, suggesting their potential importance as a source of nitrogen. A classical study showed that putrescine and spermidine could be a substitute for inorganic nitrogen for the growth of explants from the Jerusalem artichoke tubers [19]. The interplay among polyamines and nitrogen could be a key issue in the plant growth. A review by Paschalidis et al. [20] focuses on how polyamines and nitrogen regulate plant growth, especially under stress conditions. The authors also put focus on the possibility of crop improvement by genetically altering polyamine homeostasis. A paper by Anwar et al. [21] provides evidence that tomato fruit architecture can be altered by the transgenic expression of yeast spermidine synthase gene. Previous studies by these authors have shown that the increased polyamine content in transgenic tomato fruits enhances nutritional quality, juice quality, and shelf life [22,23]. The obovoid phenotype observed in tomato fruits with higher content of polyamines is shown to be associated with increased expression of fruit shape-related and cell division genes [21].

A study by Labarrere et al. [24] gives insight into a different aspect of polyamine research. The authors investigated two metabolite classes, amines and quercetins, belonging to the flavonoid family, in three *Ranunculus* species native to Kerguelen Islands. The data suggest functional redundancy and versatility of these metabolites within species, which may be attributable to adaptation to unknown past environments or neutral microevolutionary differentiation among populations.

In summary, this Special Issue contains six research articles and three reviews. I believe that these papers will further stimulate research in this exciting filed of plant biology and make readers aware of the link between their own work and topics related to polyamines.

Acknowledgments: I would like to thank all colleagues that contributed to this Special Issue and the editorial office for their helpful support during the compilation of this issue.

Conflicts of Interest: The author declares no conflict of interest.

References

1. Alcázar, R.; Tiburcio, A.F. Plant polyamines in stress and development: An emerging area of research in plant sciences. *Front. Plant Sci.* **2014**, *5*, 319. [CrossRef] [PubMed]
2. Masson, P.H.; Takahashi, T.; Angelini, R. Editorial: Molecular mechanisms underlying polyamine functions in plants. *Front. Plant Sci.* **2017**, *8*, 14. [CrossRef] [PubMed]
3. Alcázar, R.; Fortes, A.M.; Tiburcio, A.F. Editorial: Polyamines in plant biotechnology, food nutrition and human health. *Front. Plant Sci.* **2020**, *11*, 120. [CrossRef] [PubMed]
4. Takano, A.; Kakehi, J.I.; Takahashi, T. Thermospermine is not a minor polyamine in the plant kingdom. *Plant Cell Physiol.* **2012**, *53*, 606–616. [CrossRef]
5. Knott, J.M.; Römer, P.; Sumper, M. Putative spermine synthases from *Thalassiosira pseudonana* and *Arabidopsis thaliana* synthesize thermospermine rather than spermine. *FEBS Lett.* **2007**, *581*, 3081–3086. [CrossRef]
6. Kim, D.W.; Watanabe, K.; Murayama, C.; Izawa, S.; Niitsu, M.; Michael, A.J.; Berberich, T.; Kusano, T. Polyamine oxidase 5 regulates Arabidopsis growth through thermospermine oxidase activity. *Plant Physiol.* **2014**, *165*, 1575–1590. [CrossRef]
7. Alabdallah, O.; Ahou, A.; Mancuso, N.; Pompili, V.; Macone, A.; Pashkoulov, D.; Stano, P.; Cona, A.; Angelini, R.; Tavladoraki, P. The Arabidopsis polyamine oxidase/dehydrogenase 5 interferes with cytokinin and auxin signaling pathways to control xylem differentiation. *J. Exp. Bot.* **2017**, *68*, 997–1012. [CrossRef]
8. Zarza, X.; Atanasov, K.E.; Marco, F.; Arbona, V.; Carrasco, P.; Kopka, J.; Fotopoulos, V.; Munnik, T.; Gómez-Cadenas, A.; Tiburcio, A.F.; et al. *Polyamine Oxidase 5* loss-of-function mutations in *Arabidopsis thaliana* trigger metabolic and transcriptional reprogramming and promote salt stress tolerance. *Plant Cell Environ.* **2017**, *40*, 527–542. [CrossRef]
9. Sagor, G.H.M.; Kusano, T.; Berberich, T. Polyamine oxidase from Selaginella lepidophylla (SelPAO5) can replace AtPAO5 in Arabidopsis through converting thermospermine to norspermidine instead to spermidine. *Plants* **2019**, *8*, 99. [CrossRef]
10. Miyamoto, M.; Shimao, S.; Tong, W.; Motose, H.; Takahashi, T. Effect of thermospermine on the growth and expression of polyamine-related genes in rice seedlings. *Plants* **2019**, *8*, 269. [CrossRef]
11. Poidevin, L.; Unal, D.; Belda-Palazón, B.; Ferrando, A. Polyamines as quality control metabolites operating at the post-transcriptional level. *Plants* **2019**, *8*, 109. [CrossRef]
12. Yu, Z.; Jia, D.; Liu, T. Polyamine oxidases play various roles in plant development and abiotic stress tolerance. *Plants* **2019**, *8*, 184. [CrossRef]
13. Toumi, I.; Pagoulatou, M.G.; Margaritopoulou, T.; Milioni, D.; Roubelakis-Angelakis, K.A. Genetically modified heat shock protein90s and polyamine oxidases in *Arabidopsis* reveal their interaction under heat stress affecting polyamine acetylation, oxidation and homeostasis of reactive oxygen species. *Plants* **2019**, *8*, 323. [CrossRef]
14. Planas-Portell, J.; Gallart, M.; Tiburcio, A.F.; Altabella, T. Copper-containing amine oxidases contribute to terminal polyamine oxidation in peroxisomes and apoplast of *Arabidopsis thaliana*. *BMC Plant Biol.* **2013**, *13*, 109. [CrossRef] [PubMed]
15. Tavladoraki, P.; Cona, A.; Federico, R.; Tempera, G.; Viceconte, N.; Saccoccio, S.; Battaglia, V.; Toninello, A.; Agostinelli, E. Polyamine catabolism: Target for antiproliferative therapies in animals and stress tolerance strategies in plants. *Amino Acids* **2012**, *42*, 411–426. [CrossRef]
16. Cona, A.; Rea, G.; Angelini, R.; Federico, R.; Tavladoraki, P. Functions of amine oxidases in plant development and defense. *Trends Plant Sci.* **2006**, *11*, 80–88. [CrossRef] [PubMed]
17. Fraudentali, I.; Ghuge, S.A.; Carucci, A.; Tavladoraki, P.; Angelini, R.; Cona, A.; Rodrigues-Pousada, R.A. The copper amine oxidase AtCuAOδ participates in abscisic acid-induced stomatal closure in Arabidopsis. *Plants* **2019**, *8*, 183. [CrossRef] [PubMed]
18. Fraudentali, I.; Rodrigues-Pousada, R.A.; Tavladoraki, P.; Angelini, R.; Cona, A. Leaf-wounding long-distance signaling targets AtCuAOβ leading to root phenotypic plasticity. *Plants* **2020**, *9*, 249. [CrossRef]
19. Bagni, N.; Calzoni, G.L.; Speranza, A. Polyamines as sole nitrogen sources for *Helianthus tuberosus* explants in vitro. *New Phytol.* **1978**, *80*, 317–323. [CrossRef]
20. Paschalidis, K.; Tsaniklidis, G.; Wang, B.Q.; Delis, C.; Trantas, E.; Loulakakis, K.; Makky, M.; Sarris, P.F.; Ververidis, F.; Liu, J.H. The interplay among polyamines and nitrogen in plant stress responses. *Plants* **2019**, *8*, 315. [CrossRef]

21. Anwar, R.; Fatima, S.; Mattoo, A.K.; Handa, A.K. Fruit architecture in polyamine-rich tomato germplasm is determined via a medley of cell cycle, cell expansion, and fruit shape genes. *Plants* **2019**, *8*, 387. [CrossRef] [PubMed]

22. Mehta, R.A.; Cassol, T.; Li, N.; Ali, N.; Handa, A.K.; Mattoo, A.K. Engineered polyamine accumulation in tomato enhances phytonutrient content, juice quality, and vine life. *Nat. Biotechnol.* **2002**, *20*, 613–618. [CrossRef] [PubMed]

23. Nambeesan, S.; Datsenka, T.; Ferruzzi, M.G.; Malladi, A.; Mattoo, A.K.; Handa, A.K. Overexpression of yeast spermidine synthase impacts ripening, senescence and decay symptoms in tomato. *Plant J.* **2010**, *63*, 836–847. [CrossRef] [PubMed]

24. Labarrere, B.; Prinzing, A.; Dorey, T.; Chesneau, E.; Hennion, F. Variations of secondary metabolites among natural populations of sub-antarctic *Ranunculus* species suggest functional redundancy and versatility. *Plants* **2019**, *8*, 234. [CrossRef] [PubMed]

Article

A Polyamine Oxidase from *Selaginella lepidophylla* (SelPAO5) can Replace AtPAO5 in *Arabidopsis* through Converting Thermospermine to Norspermidine instead to Spermidine

G. H. M. Sagor [1], Tomonobu Kusano [2] and Thomas Berberich [3],*

[1]   Department of Genetics and Plant Breeding, Faculty of Agriculture, Bangladesh Agricultural University, Mymensingh 2202, Bangladesh; sagorgpb@gmail.com

[2]   Graduate School of Life Sciences, Tohoku University, 2-1-1 Katahira, Aoba, Sendai 980-8577, Japan; kusano@ige.tohoku.ac.jp

[3]   Laboratory Center, Senckenberg Biodiversity and Climate Research Center, Georg-Voigt-Str. 14-16, D-60325 Frankfurt am Main, Germany

*   Correspondence: tberberich@senckenberg.de; Tel.: +49-69-7542-1843

Received: 11 February 2019; Accepted: 11 April 2019; Published: 15 April 2019

Abstract: Of the five polyamine oxidases in *Arabidopsis thaliana*, AtPAO5 has a substrate preference for the tetraamine thermospermine (T-Spm) which is converted to triamine spermidine (Spd) in a back-conversion reaction in vitro. A homologue of AtPAO5 from the lycophyte *Selaginella lepidophylla* (SelPAO5) back-converts T-Spm to the uncommon polyamine norspermidine (NorSpd) instead of Spd. An *Atpao5* loss-of-function mutant shows a strong reduced growth phenotype when growing on a T-Spm containing medium. When SelPAO5 was expressed in the *Atpao5* mutant, T-Spm level decreased to almost normal values of wild type plants, and NorSpd was produced. Furthermore the reduced growth phenotype was cured by the expression of *SelPAO5*. Thus, a NorSpd synthesis pathway by PAO reaction and T-Spm as substrate was demonstrated in planta and the assumption that a balanced T-Spm homeostasis is needed for normal growth was strengthened.

Keywords: polyamine oxidase; norspermidine; thermospermine; *Selaginella lepidophylla*; *Arabidopsis thaliana* mutant

1. Introduction

Polyamines (PAs) are aliphatic compounds derived from amino acids with low molecular masses that are ubiquitously present in all living organisms [1,2]. Plants mainly contain the diamine putrescine (Put), the triamine spermidine (Spd), and the two tetraamines spermine (Spm) and thermospermine (T-Spm) [3–7], an isomer of Spm that was first discovered in thermophilic bacteria [8]. They are implicated in regulating various developmental processes such as embryogenesis, cell division, organogenesis, flowering, and senescence, as well as responses to abiotic and biotic stresses [9–14]. The biosynthesis of the polyamines Put, Spd, Spm, and T-Spm in plants is well elucidated [15]. "Lower" or non-vascular plants, such as bryophytes, mosses, and some eukaryotic algae, contain norspermidine (NorSpd) and norspermine (NorSpm) [16–18]. The biosynthesis of those uncommon PAs starts with 1,3-diaminopropane (DAP), which is produced by the metabolism of Spd and Spm through the action of terminal catabolism-type polyamine oxidase (PAO) [19]. The aminopropyl residue derived from decarboxylated S-adenosylmethionine is transferred to DAP by a putative aminopropyltransferase (APT) with relaxed substrate specificity, resulting in NorSpd, and subsequently, a second APT action converts NorSpd to NorSpm [19]. However, the occurrence of NorSpd and NorSpm has also been reported in alfalfa [20] and maize [21]. Catabolism of PAs in plants is executed by two kind of oxidases, copper-dependent amine oxidase (CuAO) and flavin-containing polyamine oxidase (PAO). PAO are reported to act in two different pathways, a terminal catabolic pathway and a back-conversion pathway [22]. The first characterized plant PAOs of maize and barley catalyze the terminal catabolic reactions [23–27]. They oxidize the carbon at the endo-side of the N4-nitrogen of Spm and Spd, producing N-(3-aminopropyl)-4-aminobutanal and 4-aminobutanal, respectively, and concomitantly 1,3-diaminopropane and H_2O_2 in both reactions [28,29]. A back-conversion reaction was first shown for *Arabidopsis thaliana* PAO1 that produces Spd from Spm and NorSpd from NorSpm in vitro [30]. The *Arabidopsis thaliana* gene family of PAO comprises five members named *AtPAO1–AtPAO5* with well characterized gene products that all function in the back-conversion of tetraamines to triamines and/or triamines to diamines, albeit with different substrate specificities [22]. AtPAO1 localizes in the cytoplasm and oxidizes Spm, T-Spm, and NorSpm, but not Spd [30], while AtPAO2, AtPAO3, and AtPAO4 localize in peroxisomes [31,32]. AtPAO2 and AtPAO3 convert Spm to Put via Spd, whereas AtPAO4 produces less Put from Spm, which is explained by the very low affinity for Spd [33]. AtPAO5 localizes in the cytoplasm and shows a preference to convert T-Spm (or Spm) to Spd [34]. *Arabidopsis pao5* mutants contain 2-fold higher T-Spm levels exhibit aerial tissue growth retardation and growth inhibition of stems and leaves at an early stage of development after external T-Spm application [4]. These findings are in accord with observations made in *Arabidopsis* plants with mutated *acaulis5* (*ACL5*) gene encoding T-Spm synthase. In this mutant (*acl5*), T-Spm content is reduced producing a dwarf phenotype with over-proliferated xylem vessels, suggesting a role of T-Spm in xylem differentiation [4,5]. Taken together, a fine-tuned T-Spm homeostasis secured by regulation of T-Spm synthase (Acaulis5) and T-Spm oxidase (AtPAO5) activities is necessary for proper xylem development and growth. A PAO from the lycophyte *Selaginella lepidophylla* (SelPAO5) with the highest sequence identity to AtPAO5 was shown to prefer T-Spm and Spm as substrates like the *Arabidopsis* homologue, but instead back-converts T-Spm to NorSpd not to Spd [35]. Here, for further characterization, we used the SelPAO5 encoding cDNA to complement the *Arabidopsis Atpao5* mutant.

2. Results

2.1. Phylogenetic Classification of SelPAO5 and Cellular Localization

Recombinant proteins of *Arabidopsis* AtPAO5 and rice OsPAO1 both prefer Spm and T-Spm as substrates and back-convert it to Spd in vitro [34,36]. These two PAOs are considered to convert T-Spm to Spd in plants. Phylogenetic relationship of PAOs identified in the genome of *Selaginella moellendorffii* [37] and SelPAO5 of *S. lepidophylla* to PAOs of *Arabidopsis* and rice is shown in Figure 1. PAO6 and PAO7 of *S. moellendorffii* (SmPAO6 and SmPAO7) are members of the clade III plant PAOs that comprise AtPAO5 and OsPAO1. SelPAO5 of *S. lepidophylla* belongs to this clade and is the homologue to SmPAO6 and SmPAO7.

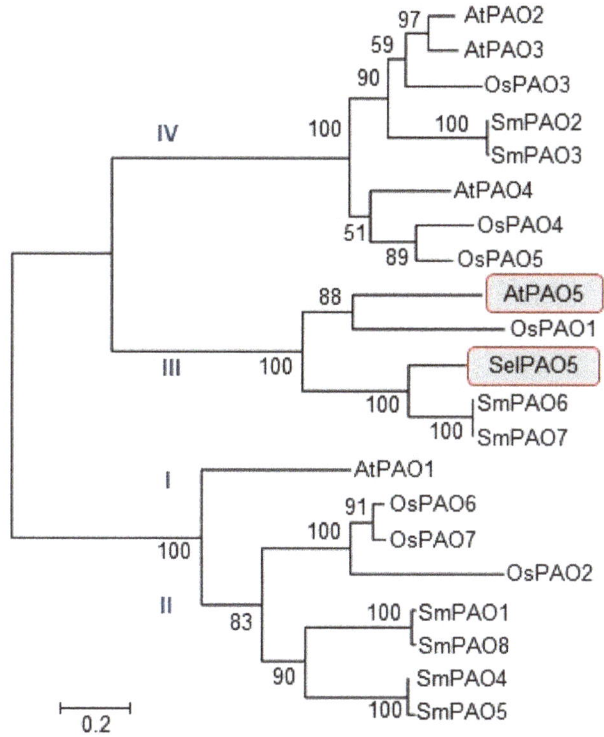

Figure 1. Phylogenetic relationship between SelPAO5, other *Selaginella* PAOs, and selected angiosperm PAOs. The tree was made by alignment of the amino acid sequences using Molecular Evolutionary Genetics Analysis (MEGA 6.0) software [38]. Bootstrap values obtained with 1000 replicates are indicated at the nodes. The genes and accession numbers used are as follows: SelPAO5 (LC036642), SmPAO1 (XP_002965265.1), SmPAO2 (XP_002965599.1), SmPAO3 (XP_002968082.1), SmPAO4 (XP_002969966.1), SmPAO5 (XP_002981437.1), SmPAO6 (XP_002984796.1), SmPAO7 (XP_002985859.1), SmPAO8 (XP_002986593.1), OsPAO1 (NM_001050573), OsPAO2 (NM_001055782), OsPAO3 (NM_001060458), OsPAO4 (NM_001060753), OsPAO5 (NM_001060754), OsPAO6 (NM_001069545), OsPAO7 (NM_001069546), AtPAO1 (NM_121373), AtPAO2 (AF364952), AtPAO3 (AY143905), AtPAO4 (AF364953), AtPAO5 (AK118203).

2.2. SelPAO5 Complementation of Atpao5-2 Mutant Rescues T-Spm-Induced Growth Inhibition

For complementing the mutated *AtPAO5* gene in *Atpao5-2* with *SelPAO5*, the cDNA of *SelPAO5* was introduced into *Atpao5-2*. *Atpao5-2* plants and *Atpao5-2* plants transformed with empty vector displayed reduced growth on T-Spm containing medium compared to wild type plants (Figure 2A), while *Atpao5-2* plants transformed with the vector containing *SelPAO5* cDNA did not. Instead, they looked similar to wild type plants (Figure 2A). For quantification of plant growth, the average fresh weight of ten seedlings each was compared (Figure 2B). While wild type and *Atpao5-2* plants expressing *SelPAO5* had an average weight of about 75 mg; the growth reduced *Atpao5-2* plants and *Atpao5-2* plants transformed with the empty vector had an average weight of 50 mg (Figure 2B). Expression of SelPAO5 was confirmed by RT-PCR with RNA samples of wild type *Arabidopsis* plants, *Atpao5-2* plants, and *Atpao5-2* plants transformed with the empty vector or the vector containing *SelPAO5* cDNA (lines S5#5, S5#11 and S5#13), respectively (Figure S1A). *AtPAO5* expression could only be detected in wild type plants but not in the *Atpao5-2* mutants. *SelPAO5* expression was confirmed in three independent *Atpao5-2* lines that were transformed with the *SelPAO5* cDNA-containing vector but not in plants that have been transformed with the empty vector only. While T-Spm had a negative effect on growth of *Atpao5-2* plants, and *Atpao5-2* plants transformed with the empty vector (Figure S1B), other polyamines, Put, Spd, and Spm, respectively, did not have such an effect (Figure S1C). In all the three lines of *Atpao5-2* expressing *SelPAO5*, no growth inhibition could be seen on T-Spm containing medium.

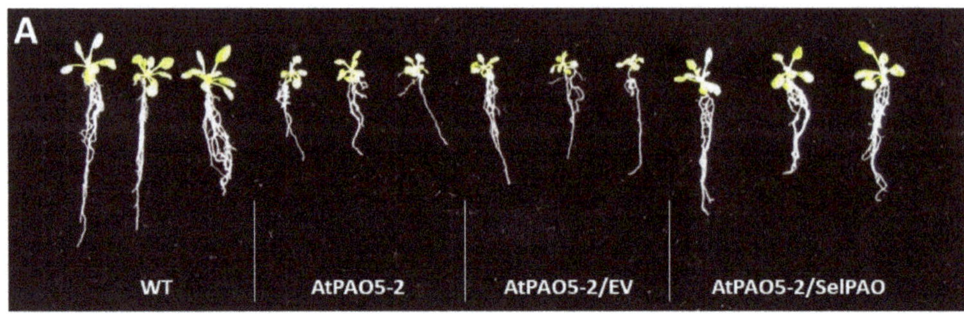

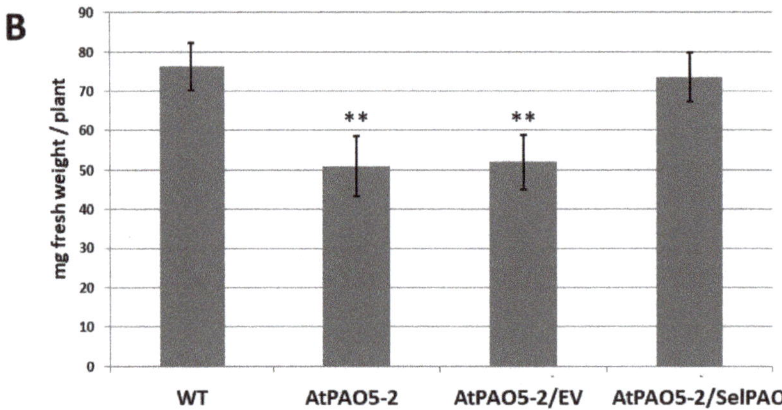

Figure 2. Recovery of thermospermine (T-Spm)-induced growth arrest in *Atpao5-2* by complementation with *SelPAO5*. Wild type plants (WT, Col-0), *Atpao5-2* mutant (AtPAO5-2), *Atpao5-2* transgenic carrying the control empty binary vector pPZP2Ha3(+) [39] (AtPAO5-2/EV), and *Atpao5-2* transgenic line S5#11 carrying the *CaMV35S*-driven *SelPAO5* (AtPAO5-2/SelPAO) were grown vertically for 24 days on half-strength Murashige and Skoog agar medium containing 5 µM T-Spm. Seedlings were carefully picked from the plates and photographed (**A**). The fresh weight of ten seedlings each was determined and the calculated mean including standard deviations displayed in a bar chart (**B**). Asterisks indicate significant differences to fresh weight of WT plants using Student's *t*-test: ** $p < 0.01$.

2.3. SelPAO Produces NorSpd in Arabidopsis Plants

Polyamine patterns in the *Atpao5-2* mutant expressing *SelPAO5* were compared to that of wild type *Arabidopsis* (Col-0) by HPLC analysis (Figure 3). In Col-0 plants, the major plant PAs Put, Spd, T-Spm, and Spm were detected but not NorSpd. The *Atpao5-2* mutant expressing *SelPAO5* (*pao5-2/SelPAO* OX) contained NorSpd in addition to the four other PAs. Quantification of PAs revealed that three lines of *Atpao5-2* mutant expressing *SelPAO5* (S5#5, S5#11, and S5#13) contained more Put (10–12 nmol/gFW) than wild type *Arabidopsis* plants, *Atpao5-2* plants, and *Atpao5-2* plants transformed with the empty vector (7–8 nmol/gFW, Figure 4A). Spd levels were similar (~50 nmol/gFW) among these plants (Figure 4B). T-Spm content in *Atpao5-2* plants and *Atpao5-2* plants transformed with the empty vector were higher (~8 nmol/gFW) than in wild type and *Atpao5-2* plants expressing *SelPAO5* (~5 nmol/gFW). The Spm content did not vary much (10–15 nmol/gFW) within the plants tested (Figure 4D). NorSpd was only detected in *Atpao5-2* plants expressing *SelPAO5* (5-9 nmol/gFW, Figure 4E). The data show that

SelPAO5 produces NorSpd when expressed in the *Atpao5-2* background. Furthermore, since the level of T-Spm drops in the *SelPAO5* expressing *Atpao5-2* mutant compared to untransformed *Atpao5-2* and *Atpao5-2* transformed with the empty vector, and the Spm contents stay almost same, it can be assumed that SelPAO5 converts T-Spm to NorSpd in planta.

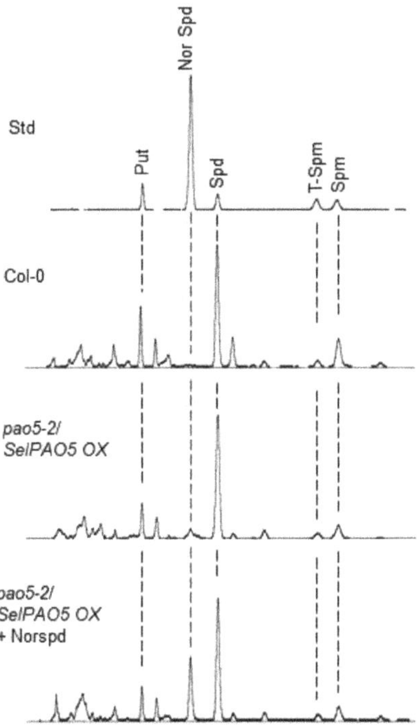

Figure 3. Chromatograms of HPLC analysis of polyamine patterns from *Arabidopsis* wild type plants (Col-0) and *Atpao5-2* mutant expressing *SelPAO5*, respectively. Std = chromatogram of polyamine standards.

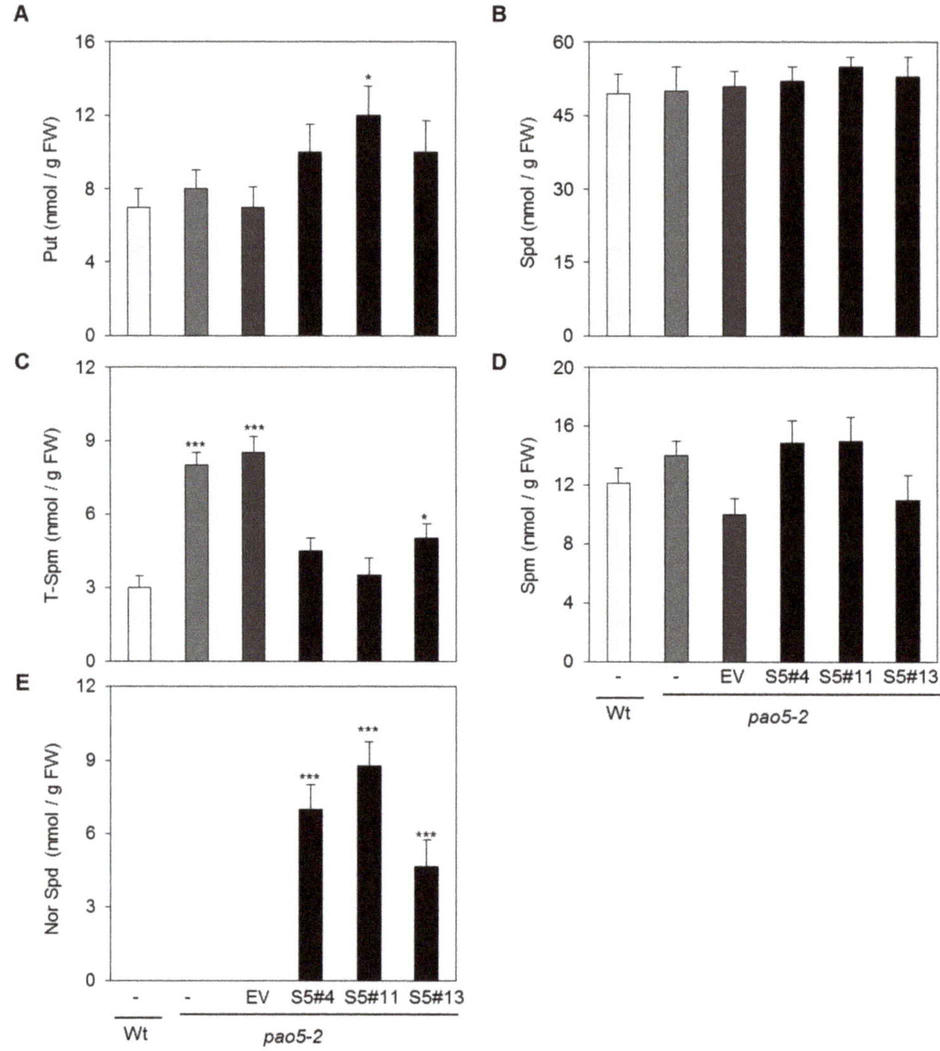

Figure 4. Polyamine content analysis of WT and pao5-2 transgenics under physiological conditions. (**A**) Put; (**B**) Spd; (**C**) T-Spm; (**D**) Spm; (**E**) NorSpd. Plant samples were: WT, *Atpao5-2*, control transgenic EV, and three transgenic lines S5#4, S5#11, and S5#13. *, **, *** indicate significance at a 5%, 1%, and 0.1% level of significance, respectively.

3. Discussion

A recent phylogenetic analysis using a plant PAO protein sequence database identified four subfamilies: three subfamilies comprising PAOs with back conversion activity named PAO back conversion 1–3 (PAObc1, PAObc2, PAObc3), and one subfamily formed by terminal catabolism PAOs (subfamily PAOtc) [40]. PAObc1 was present on every lineage in the survey, pointing out important roles of back conversion-type PAOs in plants. PAObc2 was exclusively present in vascular plants, supporting the idea that T-Spm oxidase activity plays an important role in the development of the vascular system [34,40]. *Arabidopsis* AtPAO5 and rice OsPAO1 belong to this subfamily. Based on phylogenetic relationship, polyamine oxidase SelPAO5 of *Selaginella lepidophylla* is an orthologue of

Arabidopsis AtPAO5 and rice OsPAO1 which both convert Spm and T-Spm to Spd in a back-conversion reaction. Therefore, it was expected that SelPAO5 produces Spd when using T-Spm as a substrate. However, in a previous work we could show that the recombinant SelPAO5 protein produces NorSpd in vitro [35]. To further characterize SelPAO5, we wanted to answer the questions i) does SelPAO5 convert Spm and/or T-Spm to NorSpd in vivo, and ii) can SelPAO5 replace AtPAO5 function and cure the reduced growth phenotype of *Atpao5-2* mutant? In the *Atpao5-2* mutant, T-Spm levels were increased, and plants showed a reduced growth phenotype [34]. A reduced growth phenotype was also observed in the *Arabidopsis Acaulis5* mutant (*acl5*) lacking T-Spm synthase activity and thus had decreased T-Spm levels [4,5,41]. Therefore, it is assumed that deviation from normal T-Spm levels, both an increase and decrease, cause reduced growth of plants [34]. A balanced homeostasis of T-Spm is necessary for normal growth. When SelPAO5 was expressed in the *Arabidopsis Atpao5-2* mutant, the T-Spm content decreased to almost normal levels of wild type plants while Spm levels did not decrease. NorSpd was only detected in the *Atpao5-2* mutant that expressed *SelPAO5*. In total, these results suggest that SelPAO5 uses T-Spm as substrate and converts it to NorSpd in a back-conversion reaction when expressed in *Arabidopsis*. Reduction of T-Spm content to almost wild type levels by SelPAO5 action also cured the growth retardation effect that is caused by increased T-Spm levels and enables normal development. Whether Spd or NorSpd is produced by the T-Spm specific PAO activity does not make a difference concerning the effect of T-Spm homeostasis. The presence of NorSpd does not seem to disturb development of *Arabidopsis*, although it is usually not detectable in this plant. NorSpd is an unusual triamine in eukaryotes, which is present in lower, single-celled eukaryotes including *Euglena*, cryptophytes, diatoms, and also in *Chlamydomonas* and *Volvox* [18,42,43], but also in Bryophytes [16], in the leguminous plant *Medicago sativa* (alfalfa) [20], and in maize [21]. A NorSpd synthesis pathway like in the Gram-negative bacterium *Vibrio cholerea* is not found in eukaryotes [7], and NorSpd synthesis in alfalfa from the precursor DAP (1,3-diaminopropane), which is a co-product of Spd oxidation by PAO, could not be demonstrated [44]. The production of NorSpd by SelPAO5 using T-Spm as a substrate is a demonstration of a NorSpd synthesis pathway in plants. The idea that T-Spm back-conversion by a PAO results in NorSpd in plants is backed by the finding that presence of homologues of the *Arabidopsis ACL5*-encoded T-Spm synthase in genomes correlates with the presence of NorSpd in the organism [7]. In the unicellular green alga *Chlamydomonas reinhardtii*, NorSpd stimulated cell division [45]. The role that NorSpd could play in higher plants is yet unknown. Further work should be done to follow how NorSpd is further metabolized in *Arabidopsis* and what kind of effect it has by making use of the *Atpao5-2* mutant expressing *SelPAO5*.

4. Materials and Methods

4.1. Plant Materials and Growth Conditions

Arabidopsis thaliana wild-type (WT) plants [accession Columbia-0 (Col-0)] and the T-DNA inserted *Atpao5-2* line (SALK_053110) [35] were used in this work. All seeds were surface sterilized by wetting with 70% ethanol for 1 min and subsequent treatment in a solution of 1% sodium hypochloride and 0.1% Tween-20 for 15 min. After extensive washing with sterile distilled water, sterilized seeds were placed onto vermiculite or on 1/2 Murashige and Skoog medium-1.5% agar plates (pH 5.6) containing 1% sucrose. Agar plates were kept upright under the angle of 75° to ground to allow plant growth on the agar surface by gravity. Growth conditions were 22 °C with a 14 h light/10 h dark photocycle.

4.2. Determination of Plant Fresh Weight

Seedlings grown for 24 days on agar surface containing 5 μM T-Spm where carefully picked with forceps and immediately weighed on a precision scale. Statistical analysis was done using MS-Excel software.

4.3. Chemicals

Put, Spd, and Spm were purchased from Nacalai-Tesque Ltd. (Kyoto, Japan). T-Spm and Nor-Spd was chemically synthesized [46]. All other analytical grade chemicals were obtained from Sigma-Aldrich Corp. (St. Louis, MO, USA), Wako Pure Chemical Industries Ltd. (Osaka, Japan), and Nacalai-Tesque Ltd.

4.4. Generation of Arabidopsis pao5 T-DNA Insertion Mutant Transgenic Lines Expressing SelPAO5 ORF

The fragment encompassing the coding region of the *SelPAO5* cDNA was amplified by PCR with the primer pair listed in Table S1. It was digested with *Xba*I and *Sac*I and subcloned into the corresponding sites of the pPZP2Ha3(+) vector [39], yielding pPZP2Ha3(+)-SelPAO5. This plasmid was introduced into *Agrobacterium tumefaciens* strain GV3101, and the *Agrobacterium* transformant then introduced into *pao5-2* plants using the floral dip method [47]. The resulting seeds were selected on MS agar medium containing 25 mg/mL hygromycin (hyg) and 50 mg/mL carbenicillin. T_2 seeds, obtained from self-fertilization of primary transformants, were surface-sterilized and grown on hyg-containing plates. Seedlings showing a 3:1 (resistant:sensitive) segregation ratio were selected to produce homozygous (hyg^R/hyg^R) T_3 lines that were used for further study.

4.5. RT-PCR Analysis

Total RNA was extracted from whole aerial parts of two-week-old *Arabidopsis* seedlings using Sepasol-RNA I Super (Nacalai-Tesque, Kyoto, Japan). First-strand cDNA was synthesized with ReverTra Ace (Toyobo Co. Ltd., Osaka, Japan) and oligo-dT primers. Quantitative real-time RT-PCR was performed in triplicate using Fast-Start Universal SYBR Green Master (ROX; Roche Molecular Systems, Indianapolis, IN, USA) on a StepOne real-time PCR system (Thermo Fisher Scientific, Waltham, MA, USA) using the above cDNA and the primers listed in Table S1. Constitutively expressed *AtActin* (accession number, NC_008396.2) was used as an internal control for the analysis to which the amount of target mRNA was normalized.

4.6. PA Analysis by High-Performance Liquid Chromatography (HPLC)

PA analysis was performed as described previously [5]. The benzoylated PAs were analyzed with a programmable Hewlett Packard series 1200 liquid chromatograph using a reverse-phase column (4.6 × 250 mm, TSK-GEL ODS-80Ts, TOSOH, Tokyo, Japan) and detected at 254 nm. One cycle of the run consisted of a total of 60 min at a flow rate of 1 mL/min at 30 °C; i.e., 42% acetonitrile for 25 min for PA separation, increased up to 100% acetonitrile during 3 min, then 100% acetonitrile for 20 min for washing, decreased down to 42% acetonitrile during 3 min, and finally 42% acetonitrile for 9 min. Statistical analysis was done using MS-Excel software.

Supplementary Materials: The following are available online at http://www.mdpi.com/2223-7747/8/4/99/s1, Table S1. Oligonucleotide primers used in this study, Figure S1. Recovery of T-Spm-induced growth reduction in *Atpao5-2* by complementation with *SelPAO5*.

Author Contributions: Conceptualization, G.H.M.S., T.K., and T.B.; methodology, G.H.M.S., T.K., and T.B.; validation, G.H.M.S., T.K., and T.B.; writing—original draft preparation, G.H.M.S., T.K., and T.B.; writing—review and editing, G.H.M.S., T.K., and T.B.; funding acquisition, T.K.

Funding: This research was funded by the Ministry of Education, Culture, Sports, Science, and Technology of Japan (MEXT) to TK (26_04081, 15K14705). GS was supported by German Academic Exchange Service (DAAD) program Research Stays for University Academics and Scientists.

Acknowledgments: We thank N.-H. Chua, The Rockefeller University, for making plasmid vector pGFP2 available.

Conflicts of Interest: The authors declare no conflict of interest.

References

1. Tabor, C.W.; Tabor, H. Polyamines. *Annu. Rev. Biochem.* **1984**, *53*, 749–790. [CrossRef] [PubMed]
2. Cohen, S.S. *A Guide to the Polyamines*; Oxford University Press: Oxford, UK, 1998.
3. Knott, J.M.; Römer, P.; Sumper, M. Putative spermine synthases from *Thalassiosira pseudonana* and *Arabidopsis thaliana* synthesize thermospermine rather than spermine. *FEBS Lett.* **2007**, *581*, 3081–3086. [CrossRef] [PubMed]
4. Kakehi, J.; Kuwashiro, Y.; Niitsu, M.; Takahashi, T. Thermospermine is required for stem elongation in *Arabidopsis thaliana*. *Plant Cell Physiol.* **2008**, *49*, 1342–1349. [CrossRef] [PubMed]
5. Naka, Y.; Watanabe, K.; Sagor, G.H.M.; Niitsu, M.; Pillai, M.A.; Kusano, T.; Takahashi, Y. Quantitative analysis of plant polyamines including thermospermine during growth and salinity stress. *Plant Physiol. Biochem.* **2010**, *48*, 527–533. [CrossRef]
6. Takano, A.; Kakehi, J.I.; Takahashi, T. Thermospermine is not a minor polyamine in the plant kingdom. *Plant Cell Physiol.* **2012**, *53*, 606–616. [CrossRef]
7. Michael, A.J. Polyamines in eukaryotes, bacteria, and archea. *J. Biol. Chem.* **2016**, *291*, 14896–14903. [CrossRef] [PubMed]
8. Ohshima, T. Unique polyamines produced by an extreme thermophile, *Thermus thermophilus*. *Amino Acids* **2007**, *33*, 367–372. [CrossRef]
9. Groppa, M.D.; Benavides, M.P. Polyamines and abiotic stress: Recent advances. *Amino Acids* **2008**, *34*, 35–45. [CrossRef]
10. Kusano, T.; Berberich, T.; Tateda, C.; Takahashi, Y. Polyamines: Essential factors for growth and survival. *Planta* **2008**, *228*, 367–381. [CrossRef]
11. Alcázar, R.; Altabella, T.; Marco, F.; Bortolotti, C.; Reymond, M.; Koncz, C.; Carrasco, P.; Tiburcio, A.F. Polyamines: Molecules with regulatory functions in plant abiotic stress tolerance. *Planta* **2010**, *231*, 1237–1249. [CrossRef]
12. Mattoo, A.K.; Minocha, S.C.; Minocha, R.; Handa, A.K. Polyamines and cellular metabolism in plants: Transgenic approaches reveal different responses to diamine putrescine versus higher polyamines spermidine and spermine. *Amino Acids* **2010**, *38*, 405–413. [CrossRef]
13. Minocha, R.; Majumdar, R.; Minocha, S.C. Polyamines and abiotic stress in plants: A complex relationship. *Front. Plant Sci.* **2014**, *5*, 175. [CrossRef]
14. Berberich, T.; Sagor, G.H.M.; Kusano, T. Polyamines in Plant Stress Response. In *Polyamines, A Universal Molecular Nexus for Growth, Survival, and Specialized Metabolism*; Kusano, T., Suzuki, H., Eds.; Springer: Tokyo, Japan, 2015; ISBN 978-4-431-55211-6.
15. Takahashi, T.; Tong, W. Regulation and diversity of polyamine biosynthesis in plants. In *Polyamines, A Universal Molecular Nexus for Growth, Survival, and Specialized Metabolism*; Kusano, T., Suzuki, H., Eds.; Springer: Tokyo, Japan, 2015; ISBN 978-4-431-55211-6.
16. Hamana, K.; Matsuzaki, S. Distinct difference in the polyamine compositions of Bryophyta and Pteridophyta. *J. Biochem.* **1985**, *97*, 1595–1601. [CrossRef]
17. Kuehn, G.D.; Rodriguez-Garay, B.; Bagga, S.; Phillips, G.C. Novel occurrence of uncommon polyamines in higher plants. *Plant Physiol.* **1990**, *94*, 855–857. [CrossRef]
18. Hamana, K.; Aizaki, T.; Arai, E.; Uchikata, K.; Ohnishi, H. Distribution of norspermidine as a cellular polyamine within micro green algae including non-photosynthetic achlorophyllous Polytoma, Polytomella, Prototheca and Helicosporidium. *J. Gen. Appl. Microbiol.* **2004**, *50*, 289–295. [CrossRef]
19. Fuell, C.; Elliot, K.A.; Hanfrey, C.C.; Franceschetti, M.; Michael, A.J. Polyamine biosynthetic diversity in plants and algae. *Plant Physiol. Biochem.* **2010**, *48*, 513–520. [CrossRef]
20. Rodriguez-Garay, B.; Phillips, G.C.; Kuehn, G.D. Detection of norspermidine and norspermine in *Medicago sativa*, L. (alfalfa). *Plant Physiol.* **1989**, *89*, 525–529. [CrossRef]
21. Koc, E.C.; Bagga, S.; Songstad, D.D.; Betz, S.R.; Kuehn, G.D.; Phillips, G.C. Occurrence of uncommon polyamines in cultured tissues of maize. *In Vitro Cell Dev. Biol. Plant* **1998**, *34*, 623–631. [CrossRef]
22. Kusano, T.; Kim, D.W.; Liu, T.; Berberich, T. Polyamine catabolism in plants. In *Polyamines, A Universal Molecular Nexus for Growth, Survival, and Specialized Metabolism*; Kusano, T., Suzuki, H., Eds.; Springer: Tokyo, Japan, 2015; ISBN 978-4-431-55211-6.
23. Federico, R.; Angelini, R.; Cona, A.; Niglio, A. Polyamine oxidase bound to cell walls from *Zea mays* seedlings. *Phytochemistry* **1992**, *31*, 2955–2957. [CrossRef]

24. Tavladoraki, P.; Schininà, M.E.; Cecconi, F.; Di Agostino, S.; Manera, F.; Rea, G.; Mariottini, P.; Federico, R.; Angelini, R. Maize polyamine oxidase: Primary structure from protein and cDNA sequencing. *FEBS Lett.* **1998**, *426*, 62–66. [CrossRef]

25. Radová, A.; Sebela, M.; Galuszka, P.; Frébort, I.; Jacobsen, S.; Faulhammer, H.G.; Pec, P. Barley polyamine oxidase: Characterisation and analysis of the cofactor and the N-terminal amino acid sequence. *Phytochem. Anal.* **2001**, *12*, 166–173. [CrossRef]

26. Cervelli, M.; Cona, A.; Angelini, R.; Polticelli, F.; Federico, R.; Mariottini, P. A barley polyamine oxidase isoform with distinct structural features and subcellular localization. *Eur. J. Biochem.* **2001**, *268*, 3816–3830. [CrossRef]

27. Cervelli, M.; Di Caro, O.; Di Penta, A.; Angelini, R.; Federico, R.; Vitale, A.; Mariottini, P. A novel C-terminal sequence from barley polyamine oxidase is a vacuolar sorting signal. *Plant J.* **2004**, *40*, 410–418. [CrossRef]

28. Cona, A.; Rea, G.; Angelini, R.; Federico, R.; Tavladoraki, P. Functions of amine oxidases in plant development and defence. *Trends Plant Sci.* **2006**, *11*, 80–88. [CrossRef]

29. Angelini, R.; Cona, A.; Federico, R.; Fincato, P.; Tavladoraki, P.; Tisi, A. Plant amine oxidases "on the move": An update. *Plant Physiol. Biochem.* **2010**, *48*, 560–564. [CrossRef]

30. Tavladoraki, P.; Rossi, M.N.; Saccuti, G.; Perez-Amador, M.A.; Polticelli, F.; Angelini, R.; Federico, R. Heterologous expression and biochemical characterization of a polyamine oxidase from Arabidopsis involved in polyamine back conversion. *Plant Physiol.* **2006**, *141*, 1519–1532. [CrossRef]

31. Moschou, P.N.; Sanmartin, M.; Andriopoulou, A.H.; Rojo, E.; Sanchez-Serrano, J.J.; Roubelakis-Angelakis, K.A. Bridging the gap between plant and mammalian polyamine catabolism: A novel peroxisomal polyamine oxidase responsible for a full back-conversion pathway in Arabidopsis. *Plant Physiol.* **2008**, *147*, 1845–1857. [CrossRef]

32. Kamada-Nobusada, T.; Hayashi, M.; Fukazawa, M.; Sakakibara, H.; Nishimura, M. A putative peroxisomal polyamine oxidase AtPAO4 is involved in polyamine catabolism in *Arabidopsis thaliana*. *Plant Cell Physiol.* **2008**, *49*, 1272–1282. [CrossRef] [PubMed]

33. Fincato, P.; Moschou, P.N.; Spedaletti, V.; Tavazza, R.; Angelini, R.; Federico, R.; Roubelakis-Angelakis, K.A.; Tavladoraki, P. Functional diversity inside the Arabidopsis polyamine oxidase gene family. *J. Exp. Bot.* **2011**, *62*, 1155–1168. [CrossRef]

34. Kim, D.W.; Watanabe, K.; Murayama, C.; Izawa, S.; Niitsu, M.; Michael, A.J.; Berberich, T.; Kusano, T. Polyamine oxidase 5 regulates *Arabidopsis thaliana* growth through a thermospermine oxidase activity. *Plant Physiol.* **2014**, *165*, 1575–1590. [CrossRef] [PubMed]

35. Sagor, G.H.M.; Inoue, M.; Kim, D.W.; Kojima, S.; Niitsu, M.; Berberich, T.; Kusano, T. The polyamine oxidase from lycophyte *Selaginella lepidophylla* (SelPAO5), unlike that of angiosperms, back-converts thermospermine to norspermidine. *FEBS Lett.* **2015**, *589*, 3071–3078. [CrossRef]

36. Liu, T.; Kim, D.W.; Niitsu, M.; Berberich, T.; Kusano, T. *Oryza sativa* polyamine oxidase 1 back-converts tetraamines, spermine and thermospermine, to spermidine. *Plant Cell Rep.* **2014**, *33*, 143–151. [CrossRef]

37. Banks, J.A.; Nishiyama, T.; Hasebe, M.; Bowman, J.L.; Gribskov, M.; dePamphilis, C.; Albert, V.A.; Aono, N.; Aoyama, T.; Ambrose, B.A.; et al. The Selaginella genome identifies genetic changes associated with the evolution of vascular plants. *Science* **2012**, *332*, 960–963. [CrossRef]

38. Tamura, K.; Stecher, G.; Peterson, D.; Filipski, A.; Kumar, S. MEGA6: Molecular Evolutionary Genetics Analysis version 6.0. *Mol. Biol. Evol.* **2013**, *30*, 2725–2729. [CrossRef]

39. Fuse, T.; Sasaki, T.; Yano, M. Ti-plasmid vectors useful for functional analysis of rice genes. *Plant Biotech.* **2001**, *18*, 219–222. [CrossRef]

40. Bordenave, C.D.; Granados Mendoza, C.; Jiménez Bremont, J.F.; Gárriz, A.; Rodríguez, A.A. Defining novel plant polyamine oxidase subfamilies through molecular modeling and sequence analysis. *BMC Evol. Biol.* **2019**, *19*, 28. [CrossRef]

41. Hanzawa, Y.; Takahashi, T.; Komeda, Y. ACL5: An Arabidopsis gene required for internodal elongation after flowering. *Plant J.* **1997**, *12*, 863–874. [CrossRef]

42. Hamana, K.; Niitsu, M.; Samejima, K. Unusual polyamines in aquatic plants: The occurrence of homospermidine, norspermidine, thermospermine, norspermine, aminopropylhomospermidine, bis(aminopropyl)ethanediamine, and methylspermidine. *Can. J. Bot.* **1998**, *76*, 130–133.

43. Hamana, K.; Matsuzaki, S. Widespread occurrence of norspermidine and norspermine in eukaryotic algae. *J. Biochem.* **1982**, *91*, 1321–1328. [CrossRef]

44. Bagga, S.; Rochford, J.; Klaene, Z.; Kuehn, G.D.; Phillips, G.C. Putrescine aminopropyltransferase is responsible for biosynthesis of spermidine, spermine, and multiple uncommon polyamines in osmotic stress-tolerant Alfalfa. *Plant Physiol.* **1997**, *114*, 445–454. [CrossRef]
45. Tassoni, A.; Awad, N.; Griffiths, G. Effect of ornithine decarboxylase and norspermidine in modulating cell division in the green alga *Chlamydomonas reinhardtii*. *Plant Physiol. Biochem.* **2018**, *123*, 125–131. [CrossRef] [PubMed]
46. Niitsu, M.; Samejima, K. Synthesis of a series of linear pentaamines with three and four methylene chain intervals. *Chem. Pharm. Bull.* **1986**, *34*, 1032–1038. [CrossRef]
47. Clough, S.J.; Bent, A.F. Floral dip: A simplified method for Agrobacterium-mediated transformation of *Arabidopsis thaliana*. *Plant J.* **1998**, *16*, 735–743. [CrossRef] [PubMed]

Communication

Effect of Thermospermine on the Growth and Expression of Polyamine-Related Genes in Rice Seedlings

Minaho Miyamoto, Satoshi Shimao, Wurina Tong, Hiroyasu Motose and Taku Takahashi *

Graduate School of Natural Science and Technology, Okayama University, Okayama 700-8530, Japan
* Correspondence: perfect@cc.okayama-u.ac.jp; Tel.: +81-86-251-7858

Received: 31 May 2019; Accepted: 5 August 2019; Published: 6 August 2019

Abstract: A mutant defective in the biosynthesis of thermospermine, *acaulis5* (*acl5*), shows a dwarf phenotype with excess xylem vessels in *Arabidopsis thaliana*. Exogenous supply of thermospermine remarkably represses xylem differentiation in the root of seedlings, indicating the role of thermospermine in proper repression of xylem differentiation. However, the effect of thermospermine has rarely been investigated in other plant species. In this paper, we examined its effect on the growth and gene expression in rice seedlings. When grown with thermospermine, rice seedlings had no clearly enlarged metaxylem vessels in the root. Expression of *OsACL5* was reduced in response to thermospermine, suggesting a negative feedback control of thermospermine biosynthesis like in Arabidopsis. Unlike Arabidopsis, however, rice showed up-regulation of phloem-expressed genes, *OsHB5* and *OsYSL16*, by one-day treatment with thermospermine. Furthermore, expression of *OsPAO2* and *OsPAO6*, encoding extracellular polyamine oxidase whose orthologs are not present in Arabidopsis, was induced by both thermospermine and spermine. These results suggest that thermospermine affects the expression of a subset of genes in rice different from those affected in Arabidopsis.

Keywords: Arabidopsis; phloem; rice; spermine; thermospermine; xylem

---

1. Introduction

Thermospermine synthase is encoded by the *ACAULIS5* (*ACL5*) gene in *Arabidopsis thaliana* [1] and its loss-of-function mutant *acl5* shows a stunted growth along with excess xylem vessels [2–5]. The point-mutated *acl5* mRNA level is increased in *acl5* but is reduced by exogenous thermospermine [6], suggesting a negative feedback control of thermospermine biosynthesis. Previous studies have revealed that thermospermine plays a role in the repression of xylem differentiation through enhancing mRNA translation of *SAC51*, which encodes a basic helix-loop-helix (bHLH) protein [6,7]. The *SAC51* mRNA contains conserved upstream open-reading-frames (uORFs) in the 5′ leader region [8] and thermospermine functions in alleviating the inhibitory effect of the uORFs on the main ORF translation, although its precise mode of action remains unclear [9,10]. Expressions of *ACL5* and a member of the *SAC51* family, *SACL3*, are directly activated by bHLH heterodimers LHW-TMO5 and LHW-T5L1 in xylem precursor cells in the root [11] and these heterodimers play a key role in auxin-induced xylem formation [12]. Because SAC51 and SACL3 in turn compete with TMO5 or T5L1 to heterodimerize with LHW [11,13], thermospermine appears to be a part of the negative feedback regulation of auxin-induced xylem formation and thermospermine biosynthesis [14–16]. These results have been obtained solely from studies in Arabidopsis and, except for a study that reported on the negative feedback regulation of thermospermine homeostasis in poplar xylem tissues [17] and that showed the dwarf phenotype of cotton plants by silencing of *GhACL5* [18], the role of thermospermine in plant development has rarely been investigated in other plant species so far.

On the other hand, genes involved in polyamine biosynthesis and catabolism have been increasingly characterized in rice. Phylogenetic relationships based on recent studies together with the information of rice and Arabidopsis genome databases reveal that the rice genome has one gene for spermidine synthase, *OsSPDS*, one gene for thermospermine synthase, *OsACL5*, and two genes for spermine synthase, *OsSPMS1* and *OsSPMS2* (Figure 1A). Six putative genes for *S*-adenosylmethinine decarboxylase (SAMDC/AdoMetDC) are present in rice [19–21], among which two encode very short deduced polypeptides and may represent pseudo-genes. In Arabidopsis, *SAMDC4* has been suggested to be tightly involved in the biosynthesis of thermospermine because its expression is limited to vasculature and decreased by thermospermine and its loss-of-function mutant, *bud2*, shows a dwarf phenotype similar to that of *acl5* [22,23]. The orthologous gene in rice may be *OsSAMDC3* (Figure 1B). There are seven genes for polyamine oxidase (PAO) in rice [24–26]. *OsPAO2*, *OsPAO6*, and *OsPAO7*, may encode apoplastic enzymes catalyzing the terminal catabolism reaction [27] while their closest homolog in Arabidopsis, *AtPAO1*, encodes a cytoplasmic enzyme catalyzing the back-conversion of thermospermine or spermine to spermidine [28]. In Arabidopsis, *PAO5* is specifically expressed in vasculature and may be preferentially involved in the degradation of thermospermine [29–32]. The orthologous gene in rice appears to be *OsPAO1* (Figure 1C) [25,33]. According to the above-mentioned references and molecular phylogenetic trees, the pathways that are or may be catalyzed by these gene products in rice and Arabidopsis are summarized in Figure 1D. Expression of these rice genes, however, remains to be investigated in terms of the response to thermospermine.

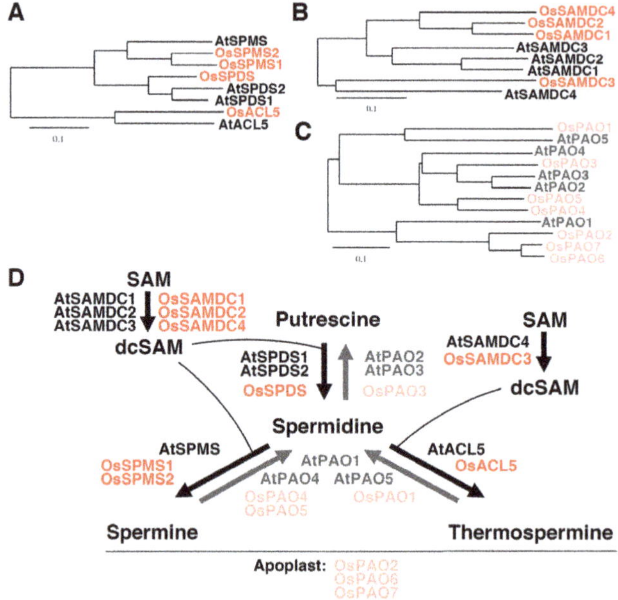

Figure 1. Polyamine biosynthetic and catabolic genes in Arabidopsis and rice. (**A**) Phylogenetic tree of spermidine synthase (SPDS), spermine synthase (SPMS), and thermospermine synthase (ACL5) isoforms. (**B**) Phylogenetic tree of *S*-adenosylmethionine decarboxylase (SAMDC) isoforms. (**C**) Phylogenetic tree of polyamine oxidase (PAO) isoforms. All trees based on the deduced amino acid sequences were constructed using ClustalW with the neighbor-joining method in the DNA Data Bank of Japan (DDBJ) website. The scale bar indicates the number of amino acid substitutions per site. (**D**) Pathways of polyamine biosynthesis and catabolism and the name of gene products mediating each reaction. Arabidopsis proteins are shown in black and light gray and rice proteins are in red and light red.

Within this context, we here focus on the effect of thermospermine on the growth and expression of the genes related to polyamine biosynthesis, catabolism, and vascular development in rice as a model monocotyledonous species.

2. Results

2.1. Thermospermine Suppresses Xylem Vessel Expansion in the Root

When grown for four days after germination with 50 µM thermospermine, rice seedlings displayed no obvious alteration in the shoot growth compared with those with mock or 50 µM spermine (Figure 2A). However, the length of the primary root was reduced by thermospermine and also by spermine (Figure 2B), indicating that these tetraamines have an inhibitory effect on root elongation in rice. The number of crown roots was not altered by these tetraamines (Figure 2C). In Arabidopsis, exogenous supply of thermospermine severely suppresses differentiation of xylem vessels and also formation of lateral roots [16,34]. In rice, formation of lateral roots was not suppressed by thermospermine (Figure 2D). Microscopic observations of the cross section revealed that rice seedlings grown with thermospermine had no clearly enlarged metaxylem vessels in the root in contrast to those grown with spermine or with no polyamines (Figure 2E) while they showed normal vascular development in leaves (Figure S1).

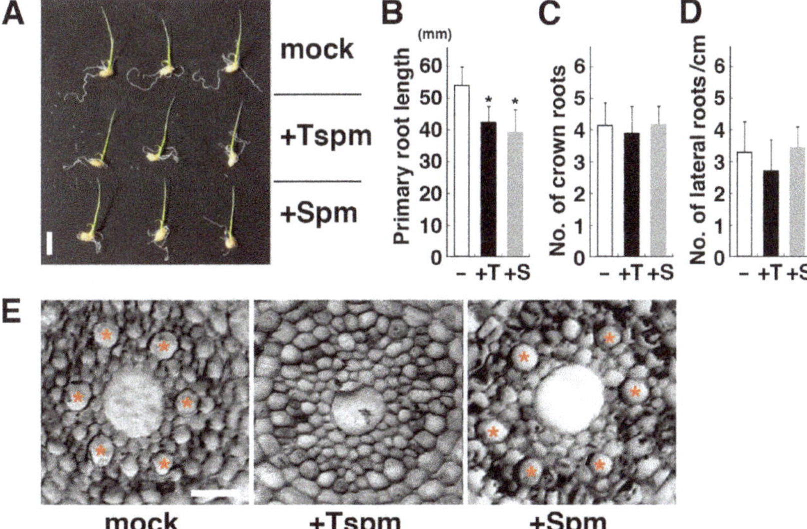

Figure 2. Effect of thermospermine and spermine on the growth of rice seedlings. (**A**) Four-day-old seedlings of rice grown in distilled water with no polyamines (mock), with 50 µM thermospermine (+Tspm), or with 50 µM spermine (+Spm). Scale bar is equivalent to 1 cm. (**B**) Length of the primary root in 4-day-old rice seedlings. (**C**) Number of crown roots in 4-day-old rice seedlings. (**D**) Density of lateral roots per length of axial root in 4-day-old rice seedlings. In (**B–D**), seedlings were grown without polyamines (−), 50 µM thermospermine (+T), or 50 µM spermine (+S). Error bars represent SE of five independent experiments with each four seedlings per treatment. Asterisks indicate significantly different values from the control (* $P < 0.05$, Student's *t*-test). (**E**) Root cross sections prepared from at a 1 cm distance from the root tip of 4-day-old rice seedling. Asterisks indicate xylem vessels with apparent diameter of more than 10 µm. Scale bar is equivalent to 20 µm.

2.2. Thermospermine Reduces OsACL5 Expression but Induces Apoplastic PAO Genes

We examined whether expression of rice genes involved in polyamine biosynthesis and catabolism are affected by exogenous thermospermine or not. We first examined the time course of the response of *OsACL5* to thermospermine in shoot and root tissues of the seedling and found that *OsACL5* expression was gradually reduced in the root during treatment with 50 μM thermospermine but not altered in the shoot (Figure 3A), suggesting that thermospermine biosynthesis in rice is also under negative feedback control at least in the root. We then focused on the effect of 24-h treatment of the seedlings with thermospermine and spermine on the expression of polyamine-related genes in the root. *OsACL5* was not responsive to spermine and *OsSPDS*, *OsSPMS1*, and *OsSPMS2* also showed no response to thermospermine and spermine (Figure 3B). Expression of *OsSAMDC2* was increased by both thermospermine and spermine, but *OsSAMDC1* and *OsSAMDC4* were not responsive to these polyamines (Figure 3C). Expression of *OsSAMDC3*, a putative ortholog of *AtSAMDC4* involved in thermospermine biosynthesis (Figure 1D), was not detected in the root. On the other hand, expressions of *OsPAO2* and, in particular, *OsPAO6*, both encoding extracellular polyamine oxidases, were drastically increased by both thermospermine and spermine while those of *OsPAO3* and *OsPAO4* were moderately increased by these tetraamines but *OsPAO5* expression was not (Figure 3D). We detected no increase in the expression of *OsPAO1*, a putative ortholog of thermospermine-catabolizing *AtPAO5*, although a previous study has reported on its induction by thermospermine [25]. Expression of *OsPAO7*, which has been shown to be an anther-specific gene [27], was not detected in the root. We confirmed that expression levels of these genes were not altered by thermospermine in the above-ground part of seedlings except for those of *PAO2* and *PAO6*, which were also increased by thermospermine (Figure S2).

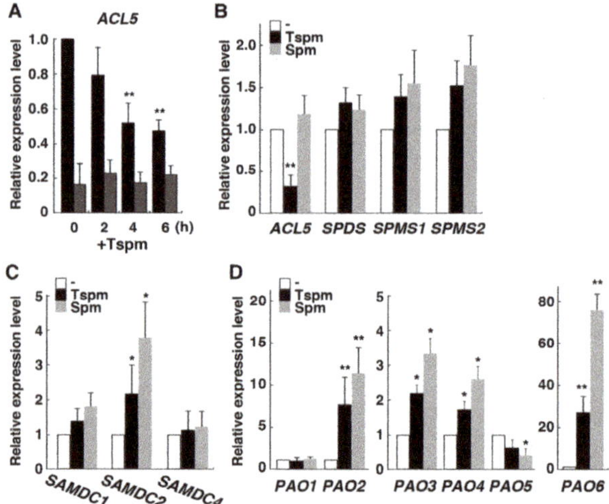

Figure 3. Effect of thermospermine and spermine on polyamine biosynthetic and catabolic genes in rice. (**A**) Time course changes in *OsACL5* expression after treatment of 4-day-old seedlings with 50 μM thermospermine (+Tspm). Black and dark gray bars represent relative expression levels in the root and those in the shoot, respectively. The level in the root before treatment is set as 1. (**B**) Expression levels of *OsACL5*, *OsSPDS*, *OsSPMS1*, and *OsSPMS2*. (**C**) Expression levels of *OsSAMDC* genes. (**D**) Expression levels of *OsPAO* genes. In (**B–D**), RNA was extracted from roots after 24-h treatment of 4-day-old seedlings with mock (white bars), 50 μM thermospermine (black bars), and 50 μM spermine (gray bars). The transcript levels in mock treated roots as represented by white bars are set as 1. Error bars represent SE of three independent experiments with each performed in duplicate. Asterisks indicate significantly different values from the control level (* $P < 0.05$, ** $P < 0.01$, Student's *t*-test).

2.3. Expression of Phloem-Specific Genes are Increased by Thermospermine

In Arabidopsis, thermospermine reduces expression of a number of genes involved in xylem differentiation [34], including all members of the Class III homeodomain leucine zipper (HD-Zip III) gene family, which is known to play a regulatory role in vascular development in Arabidopsis [35]. Then we examined the effect of thermospermine on the expression of vascular-related genes in rice roots. *OsHB3* expression was not altered by 24-h treatment with thermospermine, while *OsHB4* expression was reduced and *OsHB5* expression was increased by the same treatment (Figure 4). *OsHB3* and *OsHB4* are the closest homologs of *PHB* and *PHV* of the Arabidopsis HD-Zip III gene family and phloem-specific *OsHB5* shows the highest similarity to other HD-Zip III members, *ATHB8* and *CNA*, although expression patterns in rice are different from those in Arabidopsis [36]. We also examined expression of OsYSL16, which encodes a copper-nicotianamine transporter localized in phloem [37], and that of OsHKT1;5 encoding a sodium transporter in root xylem parenchymal cells [38]. *OsYSL16* expression was increased in response to thermospermine while *OsHKT1;5* showed no response (Figure 4). Furthermore, four genes belonging to the *SAC51* family [39] were examined. Expression of *OsSACL2* and *OsSACL3C* was markedly increased by 24-h treatment with thermospermine but *OsSACL3B* expression was reduced by thermospermine and spermine (Figure 4). In the shoot part, however, expression levels of these genes were not altered by 24-h treatment with thermospermine (Figure S2).

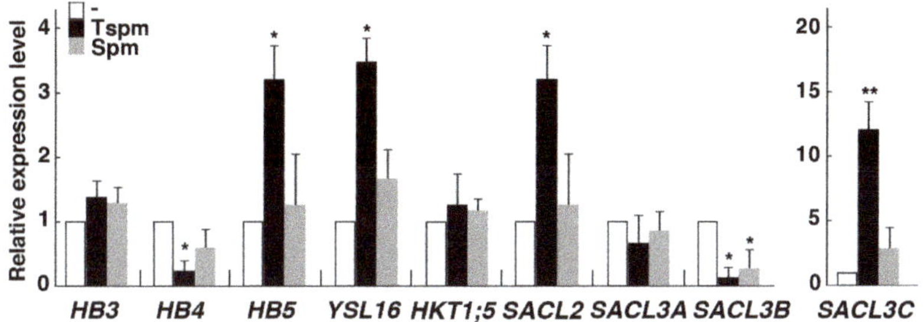

Figure 4. Effect of thermospermine and spermine on the expression of vascular-related genes in rice roots. RNA was extracted from roots sampled 24-h after treatment of 4-day-old seedlings with mock (white bars), 50 μM thermospermine (black bars), and 50 μM spermine (gray bars). The transcript levels in mock treated roots as represented by white bars are set as 1. Error bars represent SE of three independent experiments with each performed in duplicate. Asterisks indicate significantly different values from the level before treatment (* $P < 0.05$, ** $P < 0.01$, Student's *t*-test).

3. Discussion

This study provides an initial investigation of physiological and molecular effects of exogenous thermospermine in rice seedlings. We confirmed that thermospermine represses both development of metaxylem vessels and expression of *OsACL5* in the root, suggesting its functional conservation between eudicots and monocots. It may be concluded that thermospermine, whose biosynthesis is under negative feedback control of the expression of a gene for thermospermine synthase, is involved in the repression of xylem development at least in the root of angiosperms. We observed no apparent effect of thermospermine on leaf vasculature and *OsACL5* expression in the shoot. Optimal concentrations of thermospermine might not be reached at the site of action in the shoot probably because of the presence of apoplastic PAOs. Alternatively, it is also conceivable that in the rice shoot, vascular development is uncoupled from the function of thermospermine. In addition, lateral root formation, which is severely repressed by exogenous thermospermine in Arabidopsis, was not affected in rice. Development of the protoxylem, which plays a role in triggering lateral root initiation, might be less affected in rice than in Arabidopsis under our experimental condition of thermospermine treatment.

In terms of root elongation, both thermospermine and spermine were shown to be inhibitory to the growth. For rice seedlings that are normally grown in water, high concentrations of these polyamines may be generally toxic to the growth. The inhibitory effect of spermine on the growth of rice seedlings has been reported previously [40]. We found that except *OsACL5*, expressions of many polyamine-related genes were more or less increased by both thermospermine and spermine. These include *OsSAMDC2*, *OsPAO2*, *OsPAO3*, *OsPAO4*, and *OsPAO6*. There is accumulating evidence suggesting the importance of PAOs, in particular, apoplastic PAOs in stress responses [41–43]. A previous study has shown that expressions of some of rice *PAO* genes are responsive to salt stress [44]. We suggest that exogenous thermospermine and spermine may mimic the stress signal to trigger *OsPAO* gene activation. However, such a drastic increase in the expression of *PAO* genes by spermine or thermospermine has not been reported for *PAOs* in Arabidopsis and only the *AtPAO5* expression has been shown to be increased by high salt [45]. The high toxicity of exogenous polyamines in rice seedlings might be attributed to hydrogen peroxide or other products of apoplastic polyamine degradation. On the other hand, considering the function of SAMDC in providing a substrate for polyamine biosynthesis, induction of *OsSAMDC2* by polyamines seems contradictory. *OsSAMDC2* expression has also been shown to be increased by high salinity [21]. It is thus possible that, in water-grown rice seedlings, exogenous polycationic thermospermine and spermine are sensed as stress to induce stress-responsive genes including *OsSAMDC2*. How expressions of *OsPAOs* and *OsSAMDC2* are induced by polyamines or high salt remains an open question.

Expression of *OsHB5*, which was increased in response to thermospermine, is in contrast to that of *ATHB8* and *CNA* in Arabidopsis, which are reduced by thermospermine [34]. This might be related to different tissue expression patterns of HD-Zip III genes between rice and Arabidopsis. Importantly, expression of phloem-specific *OsYSL16* was also increased by thermospermine. This suggests a possibility that thermospermine is involved not only in repressing xylem development but also in promoting phloem development in rice. So far we have obtained no cytological evidence of the role of thermospermine in phloem development and it should be pursued in future work. Furthermore, among the members of the *SAC51* family, *OsSACL2* and *OsSACL3C* were up-regulated by thermospermine while *SACL3B* was down-regulated. Our recent study has shown that 5′ leader regions of *OsSACL3A* and *OsSACL3C* mRNAs can be responsive to thermospermine in terms of enhancing translation of the downstream reporter in transgenic Arabidopsis [39] while no genes that are transcriptionally and specifically up-regulated by thermospermine have been identified in Arabidopsis. Tissue-specific expression patterns of the rice *SAC51* family genes remain to be investigated.

In conclusion, our results revealed that thermospermine plays a repressive role in xylem vessel development in the root of rice seedlings and it also affects expression of a subset of genes that are different from those reduced in Arabidopsis. This difference might be largely due to the presence of apoplastic PAOs in rice, whose gene expression was found to be strongly induced by both thermospermine and spermine.

4. Materials and Methods

4.1. Chemicals

Spermine-4HCl and thermospermine-4HCl were obtained from Nakarai Chemicals (Kyoto, Japan) and Santa Cruz Biotechnology (Santa Cruz, CA, USA), respectively.

4.2. Plant Material and Growth Conditions

The rice *Oryza sativa* L. cv. Nipponbare was used throughout this work. Seeds were sown in Petri dishes containing distilled water and incubated at 22 °C on an orbital shaker at 60 rpm under 16 h light/8 h dark condition. The dishes were arranged as a single layer in a randomized complete block design. The growth experiments were repeated five times with each four seedlings per treatment.

4.3. Preparation of Sections and Microscopy

Root tissues were fixed in FAA (45% ethanol, 4% formaldehyde, 5% acetic acid), dehydrated through an ethanol series, and embedded in Technovit 7100 resin (Heraeus Kulzer, Wehrheim, Germany). Samples were sectioned into 10 μm-thick slices by a rotary microtome (RM2245, Leica Microsystems, Wetzlar, Germany) equipped with a tungsten carbide disposable blade (TC65, Leica). Sections were stained with 0.5% Toluidine blue and observed under a differential interference contrast microscope (SMZ-ZT-1, Nikon, Tokyo, Japan).

4.4. RNA Preparation and RT-PCR

Total RNA was extracted from tissues with mortar and pestle in the presence of liquid nitrogen and isolated by using NucleoSpin RNA Plant kit (Macherey-Nagel, Düren, Germany) according to the manufacture's instruction. Each RNA sample was prepared three times independently. The resulting RNA solution was treated with RNase-free DNase I (Takara, Kyoto, Japan). First-strand cDNA was synthesized using PrimeScript II 1st strand cDNA synthesis kit (Takara) with an oligo (dT) primer. Quantitative real-time PCR was performed using KAPA SYBR Fast qPCR kit (Kapa Biosystems, Woburn, MA, USA) with gene-specific primers (Tables 1 and 2) in a Thermal Cycler Dice TP760 System (Takara). *OsACT1* [46] was used as an internal control. All RT-PCR experiments were done in duplicate using three distinct cDNA preparations per sample.

4.5. Phylogenetic Analysis

Deduced protein sequences were aligned by ClustalW and phylogenetic trees were constructed using the neighbor-joining method in the DDBJ website (http://clustalw.ddbj.nig.ac.jp/index.php?lang=ja). The trees were visualized by TreeDyn (http://www.phylogeny.fr/one_task.cgi?task_type=treedyn). Accession numbers of rice genes are listed in Table 1 and those of Arabidopsis genes are AtACL5, At5g19530; *AtSPDS1*, At1g23820; *AtSPDS2*, At1g70310; *AtSPMS*, At5g53120; *AtSAMDC1*, At3g02470; *AtSAMDC2*, At5g15950; *AtSAMDC3*, At3g25570; *AtSAMDC4*, At5g18930; *AtPAO1*, At5g13700; AtPAO2, At2g43020; *AtPAO3*, AT3G59050; *AtPAO4*, AT1G65840; and AtPAO5, AT4G29720.

Table 1. List of polyamine biosynthetic and catabolic genes examined in this study and primer sequences used for RT-PCR.

Gene Name	Locus ID	MSU ID	Primer Sequence
OsSPDS	Os07g0408700	LOC_Os07g22600	F: CAACATACCCTAGTGGTGTT
			R: CTAGTTGGCCTTGGATCCAA
OsSPMS1	Os06g0528600	LOC_Os06g33710	F: CCTGAAGGGAAATATGATGC
			R: AATGACACCACTAGGATAGG
OsSPMS2	Os02g0254700	LOC_Os02g15550	F: CGACATATCCCAGTGGTGTG
			R: CAATACGCCTCTAGCTCTCT
OsACL5	Os02t0237100	LOC_Os02g14190	F: AAGAGTAGGGAGAAGTTCGA
			R: GTGTATGCTTTGACATACTTGA
OsSAMDC1	Os04g0498600	LOC_Os04g42095	F: ACTCCAACTGCGCGAAGAAG
			R: CAGCAGCAGACAAGACACCC
OsSAMDC2	Os02g0611200	LOC_Os02g39795	F: GCTTACTCCAACTGCGCGAG
			R: CGCCGAGACCGGTGGAGAGT
OsSAMDC3	Os05g0141800	LOC_Os05g04990	F: GTGGTGGACGAGAATGACCC
			R: CTAGTTGTCATGCTCATGCT

Table 1. *Cont.*

Gene Name	Locus ID	MSU ID	Primer Sequence
OsSAMDC4	Os09g0424300	LOC_Os09g25625	F: TGCTTACTCCAACTGCGCTC
			R: CATAGCCTTCAAACCCAATG
OsPAO1	Os01g0710200	LOC_Os01g51320	F: TTCCTCGGGTCATACAGCTA
			R: CTACGTGGTGTGATTCGCTC
OsPAO2	Os03g0193400	LOC_Os03g09810	F: CCCAGATTCCAATGTTCTTC
			R: GCAGAGTCAATACCTGCAAG
OsPAO3	Os04g0623300	LOC_Os04g53190	F: CTGCCGAGCCGATACATTAC
			R: CATCTCCAGCATGTCCAGCT
OsPAO4	Os04g0671200	LOC_Os04g57550	F: CCACTGAACCTACGAAGTAT
			R: ATCTCCTCGTAGGCCTTGAC
OsPAO5	Os04g0671300	LOC_Os04g57560	F: GCTACTGAACCGGTCCAGTA
			R: GGAAAAGGTCGGAGATGCCT
OsPAO6	Os09g0368200	LOC_Os09g20260	F: ACGGAGTCTGGCAGGAGTTT
			R: CGCCCTGAGCTGGTCATAC
OsPAO7	Os09g0368500	LOC_Os09g20284	F: ACGGAGTCTGGCAGGAGTTT
			R: CGCCCTGAGCTGGTCATGT

Table 2. List of other genes and their primer sequences used for RT-PCR.

Gene Name	Locus ID	MSU ID	Primer Sequence
OsHB3	Os12g0612700	LOC_Os12g41860	F: GATCATGCAGCAGGGTTTCA
			R: ATACGGTGGTGGTATTCAGG
OsHB4	Os03g0640800	LOC_Os03g43930	F: CTGCTCCCTGAAGGCTGCTC
			R: ATGACCAGTTGACGAACATG
OsHB5	Os01g0200300	LOC_Os01g10320	F: CAACATCATGGAGCAGGGGA
			R: TGTACACGCTGTTTCATGAG
OsYSL16	Os04g0542800	LOC_Os04g45900	F: CTTAACAACAGAGTGGCGGA
			R: AGAGCGCGATCTTGCCGTAG
OsHKT1;5	Os01g0307500	LOC_Os01g20160	F: CGAGGTTATCAGTGCGTATG
			R: GCATGGGTGCTTGCAGTTAG
OsSACL2	Os03g0391700	LOC_Os03g27390	F: CCCAAGATTGCCAGGCCGAG
			R: GGATCCCATCAAGAAACAACCA
OsSACL3A	Os03g0591300	LOC_Os03g39432	F: GCAAGTGTGCCAGGCCGAAT
			R: AGATCTGGGAAAGCAGGAAATG
OsSACL3B	Os02g0315600	LOC_Os02g21090	F: GTCATCGTGTGAGAGCAAG
			R: ACGTCACGGGCTTGAGAAG
OsSACL3C	Os01g0626900	LOC_Os01g43680	F: CAAGTGTGCCAGGCTGAGTA
			R: GGATCCTCTACCTGATCTGATG
OsACT1	Os03g0718100	LOC_Os03g50885	F: CTCCCCCATGCTATCCTTCG
			R: CCATCAGGAAGCTCGTAGCT

Supplementary Materials: The following are available online at http://www.mdpi.com/2223-7747/8/8/269/s1, Figure S1: Effect of thermospermine on the leaf vasculature of rice seedlings, Figure S2: Effect of thermospermine on the expression of polyamine- and vascular-related genes in the shoot of 5-day-old seedlings.

Author Contributions: Conceptualization, M.M. and T.T.; Investigation, M.M. and S.S.; Validation, M.M., S.S. and W.T.; Data analysis, M.M., W.T. and T.T.; Writing—original draft preparation, M.M. and T.T.; Writing—review and editing, H.M. and T.T.; Funding acquisition, T.T.

Funding: This work was supported in part by the Japan Society for the Promotion of Science (JSPS) Grants-in-Aid for Scientific Research (Nos. 26113516 and 19K06724) to T.T.

Conflicts of Interest: The authors declare no conflict of interest.

References

1. Knott, J.M.; Römer, P.; Sumper, M. Putative spermine synthases from *Thalassiosira pseudonana* and *Arabidopsis thaliana* synthesize thermospermine rather than spermine. *FEBS Lett.* **2007**, *581*, 3081–3086. [CrossRef] [PubMed]

2. Hanzawa, Y.; Takahashi, T.; Komeda, Y. *ACL5*, an *Arabidopsis* gene required for internodal elongation after flowering. *Plant J.* **1997**, *12*, 863–874. [CrossRef] [PubMed]

3. Hanzawa, Y.; Takahashi, T.; Michael, A.J.; Burtin, D.; Long, D.; Pineiro, M.; Coupland, G.; Komeda, Y. *ACAULIS5*, an *Arabidopsis* gene required for stem elongation, encodes a spermine synthase. *EMBO J.* **2000**, *19*, 4248–4256. [CrossRef] [PubMed]

4. Clay, N.K.; Nelson, T. *Arabidopsis* thickvein mutation affects vein thickness and organ vascularization, and resides in a provascular cell-specific spermine synthase involved in vein definition and in polar auxin transport. *Plant Physiol.* **2005**, *138*, 767–777. [CrossRef] [PubMed]

5. Muñiz, L.; Minguet, E.G.; Singh, S.K.; Pesquet, E.; Vera-Sirera, F.; Moreau-Courtois, C.L.; Carbonell, J.; Blázquez, M.A.; Tuominen, H. *ACAULIS5* controls *Arabidopsis* xylem specification through the prevention of premature cell death. *Development* **2008**, *135*, 2573–2582. [CrossRef] [PubMed]

6. Kakehi, J.I.; Kuwashiro, Y.; Niitsu, M.; Takahashi, T. Thermospermine is required for stem elongation in *Arabidopsis thaliana*. *Plant Cell Physiol.* **2008**, *49*, 1342–1349. [CrossRef] [PubMed]

7. Kakehi, J.I.; Kawano, E.; Yoshimoto, K.; Cai, Q.; Imai, A.; Takahashi, T. Mutations in ribosomal proteins, RPL4 and RACK1, suppress the phenotype of a thermospermine-deficient mutant of *Arabidopsis thaliana*. *PLoS ONE* **2015**, *27*, e0117309. [CrossRef]

8. Imai, A.; Hanzawa, Y.; Komura, M.; Yamamoto, K.T.; Komeda, Y.; Takahashi, T. The dwarf phenotype of the *Arabidopsis ACL5-1* mutant is suppressed by a mutation in an upstream ORF of a bHLH gene. *Development* **2006**, *133*, 3575–3585. [CrossRef]

9. Vera-Sirera, F.; Minguet, E.G.; Singh, S.K.; Ljung, K.; Tuominen, H.; Blázquez, M.A.; Carbonell, J. Role of polyamines in plant vascular development. *Plant Physiol. Biochem.* **2010**, *48*, 534–539. [CrossRef]

10. Takano, A.; Kakehi, J.I.; Takahashi, T. Thermospermine is not a minor polyamine in the plant kingdom. *Plant Cell Physiol.* **2012**, *53*, 606–616. [CrossRef]

11. Katayama, H.; Iwamoto, K.; Kariya, Y.; Asakawa, T.; Kan, T.; Fukuda, H.; Ohashi-Ito, K. A negative feedback loop controlling bHLH complexes is involved in vascular cell division and differentiation in the root apical meristem. *Curr. Biol.* **2015**, *25*, 3144–3150. [CrossRef] [PubMed]

12. Ohashi-Ito, K.; Saegusa, M.; Iwamoto, K.; Oda, Y.; Katayama, H.; Kojima, M.; Sakakibara, H.; Fukuda, H. A bHLH complex activates vascular cell division via cytokinin action in root apical meristem. *Curr. Biol.* **2014**, *24*, 2053–2058. [CrossRef] [PubMed]

13. Vera-Sirera, F.; De Rybel, B.; Úrbez, C.; Kouklas, E.; Pesquera, M.; Álvarez-Mahecha, J.C.; Minguet, E.G.; Tuominen, H.; Carbonell, J.; Borst, J.W.; et al. A bHLH-based feedback loop restricts vascular cell proliferation in plants. *Dev. Cell* **2015**, *35*, 432–443. [CrossRef] [PubMed]

14. Baima, S.; Forte, V.; Possenti, M.; Peñalosa, A.; Leoni, G.; Salvi, S.; Felici, B.; Ruberti, I.; Morelli, G. Negative feedback regulation of auxin signaling by ATHB8/ACL5-BUD2 transcription module. *Mol. Plant* **2014**, *7*, 1006–1025. [CrossRef] [PubMed]

15. Yoshimoto, K.; Noutoshi, Y.; Hayashi, K.; Shirasu, K.; Takahashi, T.; Motose, H. A chemical biology approach reveals an opposite action between thermospermine and auxin in xylem development in *Arabidopsis thaliana*. *Plant Cell Physiol.* **2012**, *53*, 635–645. [CrossRef] [PubMed]

16. Cai, Q.; Fukushima, H.; Yamamoto, M.; Ishii, N.; Sakamoto, T.; Kurata, T.; Motose, H.; Takahashi, T. The *SAC51* family plays a central role in thermospermine responses in *Arabidopsis*. *Plant Cell Physiol.* **2016**, *57*, 1583–1592. [CrossRef] [PubMed]

17. Milhinhos, A.; Prestele, J.; Bollhoner, B.; Matos, A.; Vera-Sirera, F.; Rambla, J.L.; Ljung, K.; Carbonell, J.; Blázquez, M.A.; Tuominen, H.; et al. Thermospermine levels are controlled by an auxin-dependent feedback loop mechanism in *Populus* xylem. *Plant J.* **2013**, *75*, 685–698. [CrossRef] [PubMed]

18. Mo, H.; Wang, X.; Zhang, Y.; Yang, J.; Ma, Z. Cotton *ACAULIS5* is involved in stem elongation and the plant defense response to *Verticillium dahliae* through thermospermine alteration. *Plant Cell Rep.* **2015**, *34*, 1975–1985. [CrossRef] [PubMed]

19. Do, P.T.; Degenkolbe, T.; Erban, A.; Heyer, A.G.; Kopka, J.; Köhl, K.I.; Hincha, D.K.; Zuther, E. Dissecting rice polyamine metabolism under controlled long-term drought stress. *PLoS ONE* **2013**, *8*, e60325. [CrossRef]

20. Chen, M.; Chen, J.J.; Fang, J.Y.; Guo, Z.F.; Lu, S.Y. Down-regulation of S-adenosylmethionine decarboxylase genes results in reduced plant length, pollen viability, and abiotic stress tolerance. *Plant Cell Tissue Org.* **2014**, *116*, 311–322. [CrossRef]

21. Saha, J.; Giri, K. Molecular phylogenomic study and the role of exogenous spermidine in the metabolic adjustment of endogenous polyamine in two rice cultivars under salt stress. *Gene* **2017**, *609*, 88–103. [CrossRef] [PubMed]

22. Ge, C.; Cui, X.; Wang, Y.; Hu, Y.; Fu, Z.; Zhang, D.; Cheng, Z.; Li, J. *BUD2*, encoding an S-adenosylmethionine decarboxylase, is required for *Arabidopsis* growth and development. *Cell Res.* **2006**, *16*, 446–456. [CrossRef] [PubMed]

23. Cui, X.; Ge, C.; Wang, R.; Wang, H.; Chen, W.; Fu, Z.; Jiang, X.; Li, J.; Wang, Y. The *BUD2* mutation affects plant architecture through altering cytokinin and auxin responses in *Arabidopsis*. *Cell Res.* **2010**, *20*, 576–586. [CrossRef] [PubMed]

24. Ono, Y.; Kim, D.W.; Watanabe, K.; Sasaki, A.; Niitsu, M.; Berberich, T.; Kusano, T.; Takahashi, Y. Constitutively and highly expressed *Oryza sativa* polyamine oxidases localize in peroxisomes and catalyze polyamine back conversion. *Amino Acids* **2012**, *42*, 867–876. [CrossRef] [PubMed]

25. Liu, T.; Kim, D.W.; Niitsu, M.; Berberich, T.; Kusano, T. *Oryza sativa* polyamine oxidase 1 back-converts tetraamines, spermine and thermospermine, to spermidine. *Plant Cell Rep.* **2014**, *33*, 143–151. [CrossRef] [PubMed]

26. Chen, B.-X.; Li, W.-Y.; Gao, Y.-T.; Chen, Z.-J.; Zhang, W.-N.; Liu, Q.-J.; Chen, Z. Involvement of polyamine oxidase-produced hydrogen peroxide during coleorhiza-limited germination of rice seeds. *Front Plant Sci.* **2016**, *7*, 1219. [CrossRef] [PubMed]

27. Liu, T.B.; Kim, D.W.; Niitsu, M.; Maeda, S.; Watanabe, M.; Kamio, Y.; Berberich, T.; Kusano, T. Polyamine oxidase 7 is a terminal catabolism-type enzyme in *Oryza sativa* and is specifically expressed in anthers. *Plant Cell Physiol.* **2014**, *55*, 1110–1122. [CrossRef] [PubMed]

28. Fincato, P.; Moschou, P.N.; Spedaletti, V.; Tavazza, R.; Angelini, R.; Federico, R.; Roubelakis-Angelakis, K.A.; Tavladoraki, P. Functional diversity inside the *Arabidopsis* polyamine oxidase gene family. *J. Exp. Bot.* **2011**, *62*, 1155–1168. [CrossRef] [PubMed]

29. Fincato, P.; Moschou, P.N.; Ahou, A.; Angelini, R.; Roubelakis-Angelakis, K.A.; Federico, R.; Tavladoraki, P. The members of *Arabidopsis thaliana PAO* gene family exhibit distinct tissue and organ-specific expression pattern during seedling growth and flower development. *Amino Acids* **2012**, *42*, 831–841. [CrossRef]

30. Ahou, A.; Martignago, D.; Alabdallah, O.; Tavazza, R.; Stano, P.; Macone, A.; Rambla, J.L.; Vera-Sirera, F.; Angelini, R.; Federico, R.; et al. A plant spermine oxidase/dehydrogenase regulated by the proteasome and polyamines. *J. Exp. Bot.* **2014**, *65*, 1585–1603. [CrossRef]

31. Kim, D.W.; Watanabe, K.; Murayama, C.; Izawa, S.; Niitsu, M.; Michael, A.J.; Berberich, T.; Kusano, T. Polyamine oxidase 5 regulates *Arabidopsis* growth through thermospermine oxidase activity. *Plant Physiol.* **2014**, *165*, 1575–1590. [CrossRef] [PubMed]

32. Alabdallah, O.; Ahou, A.; Mancuso, N.; Pompili, V.; Macone, A.; Pashkoulov, D.; Stano, P.; Cona, A.; Angelini, R.; Tavladoraki, P. The *Arabidopsis* polyamine oxidase/dehydrogenase 5 interferes with cytokinin and auxin signaling pathways to control xylem differentiation. *J. Exp. Bot.* **2017**, *68*, 997–1012. [CrossRef] [PubMed]

33. Liu, T.; Wook Kim, D.; Niitsu, M.; Berberich, T.; Kusano, T. POLYAMINE OXIDASE 1 from rice (*Oryza sativa*) is a functional ortholog of *Arabidopsis* POLYAMINE OXIDASE 5. *Plant Signal Behav.* **2014**, *9*, e29773. [CrossRef] [PubMed]

34. Tong, W.; Yoshimoto, K.; Kakehi, J.I.; Motose, H.; Niitsu, M.; Takahashi, T. Thermospermine modulates expression of auxin-related genes in *Arabidopsis*. *Front Plant Sci.* **2014**, *5*, 94. [CrossRef] [PubMed]

35. Prigge, M.J.; Otsuga, D.; Alonso, J.M.; Ecker, J.R.; Drews, G.N.; Clark, S.E. Class III homeodomain-leucine zipper gene family members have overlapping, antagonistic, and distinct roles in *Arabidopsis* development. *Plant Cell* **2005**, *17*, 61–76. [CrossRef] [PubMed]

36. Itoh, J.; Hibara, K.; Sato, Y.; Nagato, Y. Developmental role and auxin responsiveness of class III homeodomain leucine zipper gene family members in rice. *Plant Physiol.* **2008**, *147*, 1960–1975. [CrossRef] [PubMed]

37. Zheng, L.; Yamaji, N.; Yokosho, K.; Ma, J.F. YSL16 is a phloem-localized transporter of the copper-nicotianamine complex that is responsible for copper distribution in rice. *Plant Cell* **2012**, *24*, 3767–3782. [CrossRef]

38. Kobayashi, N.I.; Yamaji, N.; Yamamoto, H.; Okubo, K.; Ueno, H.; Costa, A.; Tanoi, K.; Matsumura, H.; Fujii-Kashino, M.; Horiuchi, T.; et al. OsHKT1;5 mediates Na$^+$ exclusion in the vasculature to protect leaf blades and reproductive tissues from salt toxicity in rice. *Plant J.* **2017**, *91*, 657–670. [CrossRef]

39. Ishitsuka, S.; Yamamoto, M.; Miyamoto, M.; Kuwashiro, Y.; Imai, A.; Motose, H.; Takahashi, T. Complexity and conservation of thermospermine-responsive uORFs of *SAC51* family genes in angiosperms. *Front Plant Sci.* **2019**, *10*, 564. [CrossRef]

40. Ndayiragije, A.; Lutts, S. Do exogenous polyamines have an impact on the response of a salt-sensitive rice cultivar to NaCl? *J. Plant Physiol.* **2006**, *163*, 506–516. [CrossRef]

41. Cona, A.; Rea, G.; Angelini, R.; Federico, R.; Tavladoraki, P. Functions of amine oxidases in plant development and defence. *Trends Plant Sci.* **2006**, *11*, 80–88. [CrossRef] [PubMed]

42. Wimalasekera, R.; Tebartz, F.; Scherer, G.F. Polyamines, polyamine oxidases and nitric oxide in development, abiotic and biotic stresses. *Plant Sci.* **2011**, *181*, 593–603. [CrossRef] [PubMed]

43. Moschou, P.N.; Wu, J.; Cona, A.; Tavladoraki, P.; Angelini, R.; Roubelakis-Angelakis, K.A. The polyamines and their catabolic products are significant players in the turnover of nitrogenous molecules in plants. *J. Exp. Bot.* **2012**, *63*, 5003–5015. [CrossRef] [PubMed]

44. Quinet, M.; Ndayiragije, A.; Lefevre, I.; Lambillotte, B.; Dupont, G.C.C.; Lutts, S. Putrescine differently influences the effect of salt stress on polyamine metabolism and ethylene synthesis in rice cultivars differing in salt resistance. *J. Exp. Bot.* **2010**, *61*, 2719–2733. [CrossRef] [PubMed]

45. Zarza, X.; Atanasov, K.E.; Marco, F.; Arbona, V.; Carrasco, P.; Kopka, J.; Fotopoulos, V.; Munnik, T.; Gómez-Cadenas, A.; Tiburcio, A.F.; et al. *Polyamine Oxidase 5* loss-of-function mutations in *Arabidopsis thaliana* trigger metabolic and transcriptional reprogramming and promote salt stress tolerance. *Plant Cell Environ.* **2017**, *40*, 527–542. [CrossRef] [PubMed]

46. Li, Q.F.; Sun, S.S.M.; Yuan, D.Y.; Yu, H.X.; Gu, M.H.; Liu, Q.Q. Validation of candidate reference genes for the accurate normalization of real-time quantitative RT-PCR data in rice during seed development. *Plant Mol. Biol. Rep.* **2010**, *28*, 49–57. [CrossRef]

Review

Polyamines as Quality Control Metabolites Operating at the Post-Transcriptional Level

Laetitia Poidevin [1], Dilek Unal [2], Borja Belda-Palazón [1] and Alejandro Ferrando [1,*]

[1]   Instituto de Biología Molecular y Celular de Plantas, Consejo Superior de Investigaciones Científicas-Universitat Politècnica de València, 46022 Valencia, Spain; laepoi@ibmcp.upv.es (L.P.); bbelda@ibmcp.upv.es (B.B.-P.)

[2]   Biotechnology Application and Research Center, and Department of Molecular Biology, Faculty of Science and Letter, Bilecik Seyh Edebali University, 11230 Bilecik, Turkey; dilek.unal@bilecik.edu.tr

*   Correspondence: aferrando@ibmcp.upv.es; Tel.: +34-963879931

Received: 2 April 2019; Accepted: 19 April 2019; Published: 24 April 2019

Abstract: Plant polyamines (PAs) have been assigned a large number of physiological functions with unknown molecular mechanisms in many cases. Among the most abundant and studied polyamines, two of them, namely spermidine (Spd) and thermospermine (Tspm), share some molecular functions related to quality control pathways for tightly regulated mRNAs at the level of translation. In this review, we focus on the roles of Tspm and Spd to facilitate the translation of mRNAs containing upstream ORFs (uORFs), premature stop codons, and ribosome stalling sequences that may block translation, thus preventing their degradation by quality control mechanisms such as the nonsense-mediated decay pathway and possible interactions with other mRNA quality surveillance pathways.

Keywords: polyamines; spermidine; thermospermine; nonsense-mediated decay; no-go decay; non-stop decay; quality control; translation

---

1. Introduction

The eukaryotic cell has developed sophisticated mechanisms to cope with alterations that may occur during the complex process of gene expression, which could lead to detrimental consequences. Until the final gene product is released as a functional protein, numerous sequential steps such as transcription, mRNA maturation, export from the nucleus, translation, and folding demand checkpoint controls to avoid the production of defective proteins that could compromise cell viability. The translational process is usually monitored by ribosome-associated quality control pathways able to recognize faulty events caused by halted or stalled ribosomes during the process of mRNA translation [1]. These surveillance pathways are recruited to dismantle stalled translational complexes and to degrade or recycle mRNAs, ribosomes, and aberrant nascent polypeptides to complete the clearing process of defective components and products of the translation machinery. Some of these pathways address the removal of defective polypeptides resulting from improper folding, while other specialized pathways become activated upon ribosome stalling due to defects in mRNA, translational components, or the nascent polypeptides [2]. In addition to other post-translational regulatory mechanisms, these supervision pathways participate in the course of proteostasis or protein homeostasis in the cell.

As the ribosome becomes the hub for the activity of the quality control pathways, any of the cellular metabolites or proteins linked to the ribosomal activity may impinge on the activity of the clearing pathways. In this review, we focus on the role of PAs as metabolites derived from amino acid catabolism universally present in eukaryotic cells, with essential functions linked to ribosomal activity. In particular, we compile recent information of two PAs, Tspm and Spd, whose activities in translation preventing the occurrence of ribosome stalling qualify them as quality keepers, indirectly protecting mRNAs from their degradation by preventing the activation of ribosome-associated mRNA quality control pathways.

2. Polyamines and Interactions with the Translation Machinery

PAs are small aliphatic compounds positively charged at physiological pH and are widely present in nature. PAs are enzymatically produced by the decarboxylation of the amino acids ornithine and/or arginine to provide the precursor putrescine and by the sequential addition of aminopropyl groups from decarboxylated *S*-adenosylmethionine (dcSAM) to yield higher-molecular-weight polyamines (Figure 1). The most abundant PAs in eukaryotes are the diamine putrescine (Put), the triamine Spd, and the tetraamine spermine (Spm). Plants and algae display diversity in PA biosynthesis [3], and, like some microbial thermophiles, they also contain a Spm unsymmetrical isomer named Tspm involved in plant vascular development [4].

The chemical nature of PAs has driven studies on their binding properties with respect to negatively charged macromolecules such as DNA, RNA, and phospholipids. Pioneering studies on the in vivo distribution of two highly abundant PAs (Spd and Spm) bound to macromolecules revealed that most of these two PAs in rat liver remained as a complex with RNA and to a minor extent was bound to DNA and phospholipid [5]. Previous studies had already shown that PAs could stimulate translation efficiency and fidelity in prokaryotic and eukaryotic cell-free systems, and the enhanced fidelity occurred at the level of aminoacyl-tRNA binding to the ribosome [6,7]. Moreover, by stimulating the methylation of adenine residues close to the 3′-end of the 16S rRNA, PAs were shown to be required for the efficient assembly of 30S ribosomal subunits in *E. coli* [8,9]. Refined kinetic studies with fluorescence-based assays have shown that PAs promote codon recognition by stimulating the binding of the ternary complex to the ribosome [10], in agreement with structural information obtained with photoreactive PA analogues [11–13]. Altogether, these pioneering studies strongly promote the idea that PAs are involved in translation through interactions with RNA and ribosomes [14]. In addition to their direct impact as general stimulators of translation, PAs have been shown to fine-tune the translation of specific highly regulated mRNA coding sequences by enhancing ribosomal frameshifting [15].

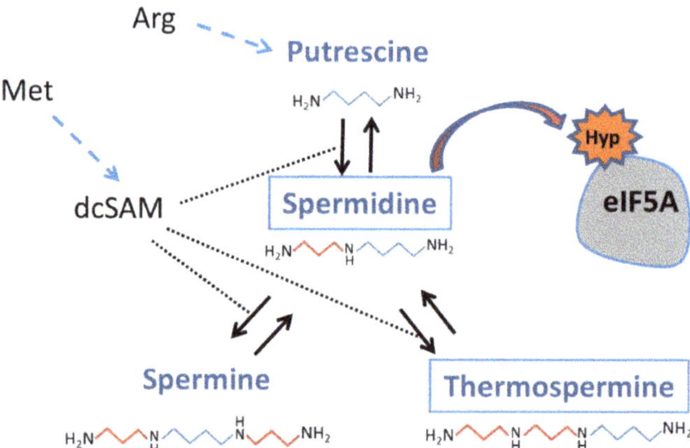

Figure 1. Polyamines derive from amino acid catabolism. In plants, the prevalent biosynthetic pathway uses arginine as a precursor for the synthesis of the diamine Put to which enzymatic addition of aminopropyl groups (in red) from decarboxylated SAM (dcSAM) yield the triamine Spd and subsequently the tetraamines Spm and Tspm. The post-translational transfer of the aminobutyl moiety (in blue) from Spd to a conserved lysine of eIF5A leads to the modified hypusine residue (Hyp) that renders an active translation factor.

3. Polyamines and mRNA Quality Control Mechanisms

Recent information on the function of two plant PAs, Tspm and Spd, has uncovered unexpected roles for PAs related to faulty translational events. These novel functions for PAs related to ribosomal activity are connected to specialized mRNA surveillance mechanisms that can detect mRNAs with defective translation and initiate their selective degradation. There are at least three known mRNA anomalies that lead to activation of the specific quality control pathways. One abundant alteration among mRNA sequences is the presence of premature termination codons (PTCs) that may lead to the production of defective proteins with dominant negative effects, in this case the surveillance machinery engaged is the nonsense-mediated decay pathway (NMD). Another type of activated system is the no-go decay pathway (NGD) that acts when ribosomes halt during elongation due to the presence of secondary mRNA structure, of rare codon repeats, or because of structural features of the nascent polypeptide that block elongation. A third route that takes care of translational failures is the non-stop decay pathway (NSD), operating on mRNAs lacking natural stop codons. In addition to the surveillance and degradation of mRNAs halted during translation, specialized complexes take care of the removal of non-functional proteins, a process described in detail elsewhere [16]. In the next subsections and sections, we briefly update each of these mRNA surveillance systems and the links to PAs respectively.

3.1. The NMD Pathway

The NMD pathway targets mRNAs carrying PTCs whose presence may generate truncated dominant-negative proteins with deleterious consequences for the cell [17–19]. The occurrence of PTCs can have multiple origins, either by non-productive alterations such as random DNA mutations or programmed genomic rearrangements, or by endogenous errors during transcription or RNA splicing [20]. PTCs can also happen in natural mRNAs as in those with upstream open reading frames (uORFs) and in transcripts of pseudogenes, transposons, and even non-coding RNAs [21–23]. Recent studies on *A. thaliana* revealed that around 1% of coding mRNAs and up to 20% of mRNA-like non-coding RNAs are targets of the NMD pathway [23]. Moreover, among the mRNA targets, a remarkable 17.4% of the multi-exon coding genes with splicing variants have been allocated to the NMD pathway, thus highlighting the importance of the alternative splicing coupled to NMD (AS-NMD) regulation [24,25]. The most important *trans*-acting factors involved in the NMD, named up-frameshift proteins (UPFs), were initially described in *S. cerevisiae* [26], and they have been shown to be highly conserved among eukaryotes [27] with only minor mechanistic differences for plants [28]. *A. thaliana upf* mutants defective in the NMD pathway have uncovered important functions for this pathway in development and stress responses [29–32]. In addition to the UPF proteins, accessory factors participate in the sequential steps of PTC detection, tagging, and destruction by endo- and exo-ribonucleases (exosome and Xnr1), all of them studied and described elsewhere with great detail [33,34]. Although the comprehensive mechanisms of NMD target recognition are still a matter of debate, there is a consensus on the participation of canonical translation termination factors eRF1 and eRF3 in the NMD process, revealing a new unexpected hypothesis as to whether this pathway not only watches for the presence of PTCs but also acts as a quality-control pathway that continuously identifies errors of premature termination during the translation process [35].

3.2. The NGD Pathway

This surveillance pathway detects ribosomes that have been stalled during translation elongation because of specific features of nascent polypeptides, secondary mRNA structures that block ribosome progression, or rare codon repeats that lead to prolonged vacancy of the A-site of the ribosome. It counts with specialized protein factors described initially in yeast and named Dom34 and Hsb1 (Dom34:Hsb1) and later in mammals and known as Pelota and HSb1-like (PELO:HSB1L) [36]. Both proteins are highly conserved and present also in plants [37,38]. The complex Dom34:Hbs1 is structurally similar to termination factors eRF1:eRF3 and in fact they do bind the vacant A-site of the ribosome to promote dissociation of the aberrant elongation complex, induce mRNA endonucleolytic cleavage, and facilitate ribosome recycling [39]. After the cleavage, the resulting mRNA fragments are subsequently degraded by the exosome and Xnr1 exonucleases like in the NMD pathway. Several questions remain to be elucidated for this pathway. First, mRNA cleavage can occur in the absence of Dom34:Hbs1 and occurs upstream of the stalling site [40]. Second, no elongation block leads to activation of the NGD, since (i) cycloheximide treatment stabilizes mRNAs and inhibits their decapping [41] and (ii) in other cases ribosomal pauses provide biological functions [42]. Therefore, the cell machinery must distinguish between useful ribosome elongation pauses or unwanted ribosomal arrests that require activation of the NGD. In this sense the degradation machinery may perceive whether the A-site is unoccupied or not and for how long this aberrant situation persists since other pathways such as translational frameshifting may be favored. An attractive model based on recent data suggests that sensing ribosome collisions trigger the endonucleolytic cleavage upstream of the stall site [43], though the identity of such endonuclease is still unknown.

3.3. The NSD Pathway

The NSD pathway degrades mRNAs that lack in-frame stop codons, either because of premature polyadenylation leading to non-stop mRNAs with poly(A) or by endonucleolytic cleavage within the open reading frame producing non-stop mRNAs lacking poly(A) [40,44]. In the case of poly(A) translation, the presence of encoded poly-lysine causes ribosome stalling probably due to interactions with negatively charged rRNA in the exit tunnel, although the nature of the poly(A) sequence also leads to a ribosome sliding behavior [45,46]. The activity of this pathway was first shown in yeast and mammals [47,48], and it has been recently shown to be operative in higher plants [49]. This pathway shares components with the NGD since a key player that was found in yeast (Ski7), structurally related to Hsb1 and eRF3, is not present in other eukaryotes. Therefore, organisms lacking Ski7 rely on Dom34/Hbs1 proteins that work for both the NGD and the NSD [40].

4. Thermospermine and uORF-Dependent Translational Inhibition: Is There A Connection to the NMD Pathway?

Tspm was first discovered in thermophilic bacteria [50], and it was later shown to be widely distributed in the plant kingdom but absent in animals and fungi [3]. Tspm, a structural isomer of Spm, is converted from Spd by the activity of the enzyme thermospermine synthase (ACL5) identified in *A. thaliana* after the isolation of a mutant deficient in Tspm named *acaulis 5* (*acl5*) [51]. The first description misidentified the enzyme encoded by *ACL5* as a spermine synthase since the analytical procedures used were unable to distinguish between Spm and Tspm [52], but it was later shown that ACL5-like enzymes synthesize Tspm both in diatoms and *A. thaliana* [53]. Different from other aminopropyltransferases, the ACL5 enzyme was probably acquired by a plant ancestor by means of horizontal transfer from archaea or bacteria [54]. As an example of how PAs, as plant growth regulators, are linked to other hormonal signals, Tspm biosynthesis has been shown to be regulated by the negative feedback regulation of an auxin signaling module [55,56]. The role of Tspm in plant development is highly specialized in vascular development with key functions in xylogenesis by preventing premature cell death [57–59]. Precisely the dramatic phenotype of the *acl5* mutant displaying stunted growth allowed for the genetic isolation of mutant suppressors, named *sac*, able to restore stem growth in the absence of Tspm to illuminate on its molecular functions [60–62]. Interestingly, several of the isolated suppressors mapped to ribosomal proteins or ribosome-associated proteins, thus suggesting molecular functions for Tspm related to the regulation of translation. Indeed, among the identified *SAC* genes, *SAC51* and its family of bHLH-like transcription factors (*SACL1-3* genes) have been key to elucidating downstream functions of Tspm in higher plants [63]. Although molecular details for the mechanism of action of Tspm still remain unknown, its downstream target functions through the activity of SACL proteins have been shown to rely on the interplay between heterodimers of bHLH transcription factors involved in vasculature formation [64].

One common feature of *SACL* genes is the presence of uORF in their 5'-leader sequence, with the peculiarity that several suppressor mutations map precisely in this region [60]. The isolation of the *sacl* mutants that no longer require Tspm indicates that Tspm is required to bypass the uORF-dependent translational inhibition of the *SACL* main ORF (mORF). The presence of uORFs, in particular those encoding conserved peptide sequences (CPuORFs), as is the case of the *SACL* genes, is often a feature to define substrates of the NMD pathway [18,23]. The genetic studies with *upf* mutants seem to corroborate this idea, as gene expression studies have shown the upregulation of the *SACL* genes in these mutants, suggesting that UPF proteins are required for the suppression of the *SACL* mRNAs [29–32]. It is therefore conceivable that Tspm activity in *SACL* translation helps to evade the premature termination on their uORFs, thus protecting these mRNAs from degradation by the NMD pathway (Figure 2). The precise mechanism of Tspm action to favor the translation of the mORF [65] is still an open question and whether this polyamine directly works within the ribosome or as an external factor is an active line of research nowadays. Another intriguing question to be solved is whether any other mRNA containing uORF could be considered as a Tspm client, as more than 20,000 uORFs have been identified in *A. thaliana* [66], although the restricted tissue expression pattern of ACL5 would certainly limit the number of potential Tspm targets containing uORFs.

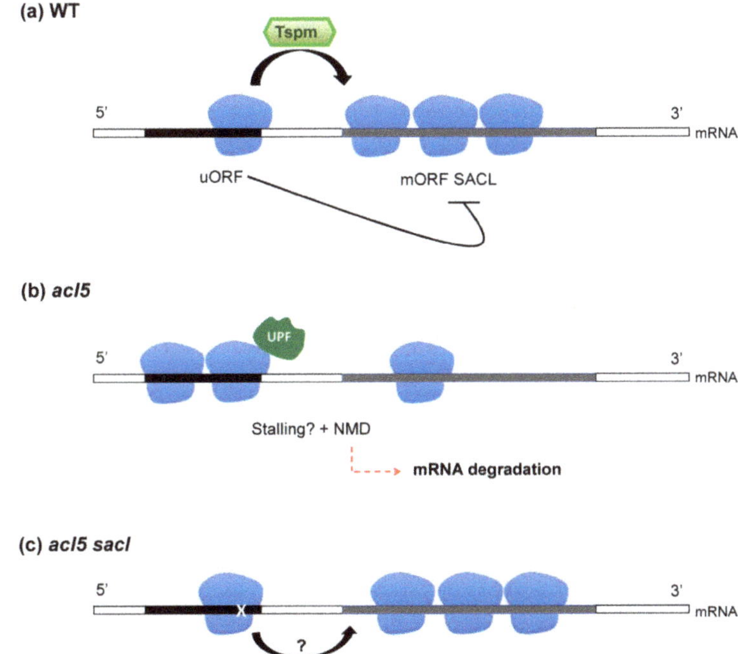

Figure 2. Thermospermine action on the translation of *SACL* genes and the connection to the NMD pathway. Under normal conditions, (**a**) the presence of Tspm, by unknown mechanistic details, prevents the blockade of translation of the main ORF (in grey) of *SACL* genes imposed by the presence of inhibitory upstream ORF (in black). *SACL* genes belong to the NMD substrate category, so it can be envisaged (**b**) that in the *acl5* mutant lacking Tspm the ribosomes may stall at the uORF and activate the NMD pathway. In the double mutants *acl5 sacl* (**c**), the presence of *cis* mutations (X in white) in the uORF of *SACLs* may alleviate ribosome stalling by changes in the peptide sequence to allow translation of the mORF or by other mechanisms to be elucidated.

5. Spermidine and Hypusination of eIF5A: How to Avoid Reversible Ribosome Pausing and Links to the NMD Pathway

The triamine Spd is a very ancient PA, and its biosynthetic enzyme spermidine synthase (SPDS) has been catalogued as one of the few protein families potentially present in the last-universal-common-ancestor (LUCA) [67]. More empirical studies have shown that Spd is an essential metabolite required for cell growth and survival in eukaryotic organisms [68–70]. One role of Spd that can explain its essential nature is its requirement as a donor for the activation by post-translational modification of the translation factor eIF5A [71]. This translation factor is the unique protein known to undergo the enzymatic modification of a conserved lysine to the unusual amino acid hypusine (Nε-(4-amino-2-hydroxybutyl)lysine) [72] by the sequential activity of first the deoxyhypusine synthase (DHS), forming the reversible intermediate deoxyhypusine, and second the deoxyhypusine hydroxylase (DOHH), catalyzing irreversibly the required hydroxylation to yield the hypusine residue [73]. As in the case of the Spd, the hypusination of translation factor eIF5A has been shown to be essential for eukaryotes' cell viability [74–77]. The eIF5A nomenclature and its assigned functions on translation have suffered chronological alterations, as it was initially named IF-M2Bα, later eIF4D, and finally eIF5A after its description as a protein that is able to stimulate the formation of methionyl-puromycin in vitro [78–80]. Hypusine discovery [81] preceded its characterization as an activator of the translation factor eIF5A and its biosynthetic origin from Spd by several years [82], but it was not until several decades later that eIF5A was shown to fulfill translation elongation rather than initiation functions in yeast [83]. To provide more complexity to the eIF5A functions carried out within the ribosome, recent studies in yeast and humans have shown that eIF5A is not only involved in elongation but also in the translation termination process [84]. Both eIF5A and DHS are present and essential in eukaryotes and archaea (named aIF5A) [85], but they are absent in eubacteria [86]. However, eubacteria possess a structurally related elongation factor named EFP that also suffers analogously though biochemically unrelated modifications essential for EFP activity [87–89]. As a difference to aIF5A and eIF5A, the knock-out of EFP is viable for the bacteria, though the mutants are more sensitive to antibiotics [90]. Structural analyses have located both EFP and eIF5A within the ribosome between the P- and the E-sites, thus supporting their role as translation elongation factors [91–93]. A major breakthrough in the field of eIF5A/EFP was the elucidation of their roles as factors required for the translation of mRNAs that encode proline rich-repeat proteins that, in the absence of these factors, lead to ribosome stalling [94–96], caused by the problematic chemical structure of proline as a rigid secondary amine being a poor peptidyl donor and receptor [97]. It is remarkable that bioinformatics analysis of poly-proline rich proteins in different eukaryotic systems from yeast to plants has suggested the idea of a parallel specialization of the Spd-eIF5A pathway with the functional organization of proline rich-repeat proteins [98]. This information has facilitated the allocation of molecular mechanisms to Spd by means of its role on the eIF5A hypusination, such as its role in yeast fertility by promoting the translation of formin, a poly-proline rich containing protein, involved in actin cytoskeleton remodeling during yeast mating [99]. The hypusination pathway in higher plants has been involved in the response to both biotic and abiotic stress conditions as well as in plant developmental processes [98,100–103] though the potential targets at the level of mRNA translation remain to be identified.

The way eIF5A/EFP resolves problematic poly-proline ribosome pauses is among the topics of excellent reviews [104,105] that summarize recent studies on this subject. One key aspect of the strategy used by eIF5A/EFP to alleviate ribosome pauses is related to their location within the ribosome, as they enter the ribosome through the E-site probably as a signal of slow translation elongation or termination that demands their intervention as a case of reversible pause. The structural studies favor the model that hypusine stabilizes the acceptor arm of the P-site peptidyl-tRNA at the peptidyl transferase center rather than contributing to the catalysis of the peptide-bond formation. This is in contrast with vacant A-sites generated by non-optimal codons, truncated mRNA, or nascent polypeptide features that, when sensed as an irreversible arrest, can trigger the degradation of the mRNA upon activation of the NGD pathway. As mentioned before, the ribosome-associated cell machinery uses advanced biochemical sensing systems to distinguish between a transient stall and a dead arrest that demands either reversible action or destructive/recycling intervention, respectively.

Although no evidence has been reported to date for the interaction of eIF5A with the NGD and NSD pathways, many studies have linked the functions of eIF5A to the NMD pathway. Pioneering searches for yeast mutants with increased stability of mRNA targets of the NMD pathway have uncovered eIF5A as an essential protein involved in mRNA decay [106]. This observation was later confirmed with double genetic disruption of yeast eIF5A-encoding genes complemented with temperature-sensitive human *EIF5A1* gene, showing elongated half-lives of NMD transcripts [107]. More recently, the functional connection between eIF5A activity and the NMD pathway has been extended to human cells [108]. One possible explanation for the increased stability of NMD-related targets is the unexplained functional link of eIF5A with stress-granule assembly [109]. Altogether these studies support the idea that Spd-dependent hypusination of eIF5A provides an NMD-related proofreading activity in translation although the underlying biochemical basis remains to be elucidated (Figure 3).

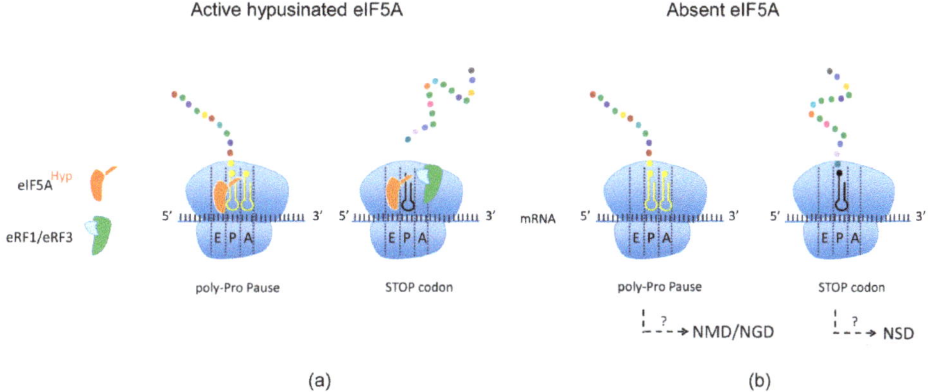

Figure 3. Spermidine-mediated hypusination of eIF5A and links to surveillance pathways. (**a**) Hypusinated eIF5A promotes elongation of stalling motives like poly-proline and stimulates eRF1-mediated peptidyl hydrolysis. (**b**) Non-functional eIF5A generates a vacant E-site during ribosome stalling at poly-proline elongation and affects termination, leading to the accumulation of NMD substrates by unknown mechanisms. Absence of eIF5A could also potentially affect NGD and NSD pathways during elongation and termination.

6. Conclusions

PAs are well-conserved metabolites with essential functions in eukaryotes. Among the different roles and locations within the cell, a considerable number of studies have linked their activities to the ribosome. This highlights vital functions for PAs, such as the essential nature of Spd for eukaryotes and the absolute requirement of Tspm for plant vascular development. The mode of action for Spd within the ribosome is to provide the aminobutyl moiety to activate the translation factor eIF5A, whereas for Tspm the available information suggests a direct and probably intimate role with the ribosome machinery. In both cases, a reduction in their levels correlates with stabilization of mRNAs belonging to the NMD pathway, thus indicating that both PAs may act as watchers for potential ribosome conflicts. In the case of Tspm, the known NMD targets belong to the group of genes with uORF sequences containing PTCs that act as translation inhibitors for the mORF. Unfortunately, the mechanistic molecular details of Tspm activity in the ribosome still need to be deciphered. On the other hand, the suppression of Spd-mediated hypusination of eIF5A leads to an accumulation of unstable transcripts, including some of the AS-NMD type, thus suggesting unknown connections of the eIF5A translation factor with the splicing machinery. Still, how hypusinated eIF5A connects to the NMD pathway awaits further studies. We cannot forget that the inactivation of eIF5A could lead to the activation of either NGD or the NSD pathways. For instance, if poly-proline induced ribosomal blocks are not rapidly resolved and lead to ribosomal collisions, this could activate NGD machinery. On the other hand, alterations in the termination process by inactive eIF5A may lead to non-stop ribosomal activity, which could activate the NSD pathway. What has been well documented is that both Tspm and Spd (through the activity of eIF5A) seem to act as molecular sentinels to prevent translation anomalies that may lead to activation of the NMD machinery and subsequent mRNA degradation. In this review, we have focused on two PAs that share similar proofreading functions on translation, but it would not be a surprise to find out that other PAs may fulfill related functions on this or related cellular processes. We anticipate that in the near future the roles of PAs in the cell in relation to ribosomal activity will be fully elucidated, thus helping to solve one of the remaining mysteries of molecular biology [110].

Author Contributions: Conceptualization and writing of the original draft: L.P. and A.F.; writing—reviewing and editing: D.U. and B.B.-P.; project administration and funding acquisition: A.F.

Funding: A.F. was funded by the Spanish Ministry of Science, Innovation and Universities, grant number BIO2015-70483-R, and B.B.-P. was funded by the Generalitat Valenciana grant, VALi+d GVA APOSTD/2017/039. D.U. was a recipient of an EMBO short-term fellowship, number STF-7308.

References

1. Graille, M.; Séraphin, B. Surveillance pathways rescuing eukaryotic ribosomes lost in translation. *Nat. Rev. Mol. Cell Biol.* **2012**, *13*, 727. [CrossRef]

2. Preissler, S.; Deuerling, E. Ribosome-associated chaperones as key players in proteostasis. *Trends Biochem. Sci.* **2012**, *37*, 274–283. [CrossRef] [PubMed]

3. Fuell, C.; Elliott, K.A.; Hanfrey, C.C.; Franceschetti, M.; Michael, A.J. Polyamine biosynthetic diversity in plants and algae. *Plant Physiol. Biochem.* **2010**, *48*, 513–520. [CrossRef]

4. Vera-Sirera, F.; Minguet, E.G.; Singh, S.K.; Ljung, K.; Tuominen, H.; Blázquez, M.A.; Carbonell, J. Role of polyamines in plant vascular development. *Plant Physiol. Biochem.* **2010**, *48*, 534–539. [CrossRef]

5. Watanabe, S.; Kusama-Eguchi, K.; Kobayashi, H.; Igarashi, K. Estimation of polyamine binding to macromolecules and ATP in bovine lymphocytes and rat liver. *J. Biol. Chem.* **1991**, *266*, 20803–20809.

6. Igarashi, K.; Sugawara, K.; Izumi, I.; Nagayama, C.; Hirose, S. Effect of Polyamines on Polyphenylalanine Synthesis by Escherichia coli and Rat-Liver Ribosomes. *Eur. J. Biochem.* **1974**, *48*, 495–502. [CrossRef] [PubMed]

7. Igarashi, K.; Hashimoto, S.; Miyake, A.; KashiwagiI, K.; Hirose, S. Increase of Fidelity of Polypeptide Synthesis by Spermidine in Eukaryotic Cell-Free Systems. *Eur. J. Biochem.* **1982**, *128*, 597–604. [CrossRef] [PubMed]

8. Echandi, G.; Algranati, I.D. Defective 30S ribosomal particles in a polyamine auxotroph of Escherichia coli. *Biochem. Biophys. Res. Commun.* **1975**, *67*, 1185–1191. [CrossRef]

9. Igarashi, K.; Kishida, K.; Hirose, S. Stimulation by polyamines of enzymatic methylation of two adjacent adenines near the 3′ end of 16S ribosomal RNA of Escherichia coli. *Biochem. Biophys. Res. Commun.* **1980**, *96*, 678–684. [CrossRef]

10. Hetrick, B.; Khade, P.K.; Lee, K.; Stephen, J.; Thomas, A.; Joseph, S. Polyamines Accelerate Codon Recognition by Transfer RNAs on the Ribosome. *Biochemistry* **2010**, *49*, 7179–7189. [CrossRef]

11. Amarantos, I.; Kalpaxis, D.L. Photoaffinity polyamines: interactions with AcPhe-tRNA free in solution or bound at the P-site of Escherichia coli ribosomes. *Nucleic Acids Res.* **2000**, *28*, 3733–3742. [CrossRef]

12. Kalpaxis, D.L.; Amarantos, I.; Zarkadis, I.K. The identification of spermine binding sites in 16S rRNA allows interpretation of the spermine effect on ribosomal 30S subunit functions. *Nucleic Acids Res.* **2002**, *30*, 2832–2843. [CrossRef]

13. Xaplanteri, M.A.; Petropoulos, A.D.; Dinos, G.P.; Kalpaxis, D.L. Localization of spermine binding sites in 23S rRNA by photoaffinity labeling: parsing the spermine contribution to ribosomal 50S subunit functions. *Nucleic Acids Res.* **2005**, *33*, 2792–2805. [CrossRef]

14. Dever, T.E.; Ivanov, I.P. Roles of polyamines in translation. *J. Biol. Chem.* **2018**. [CrossRef]

15. Ivanov, I.P.; Matsufuji, S.; Murakami, Y.; Gesteland, R.F.; Atkins, J.F. Conservation of polyamine regulation by translational frameshifting from yeast to mammals. *Embo J.* **2000**, *19*, 1907–1917. [CrossRef]

16. Brandman, O.; Hegde, R.S. Ribosome-associated protein quality control. *Nat. Struct. Mol. Biol.* **2016**, *23*, 7. [CrossRef] [PubMed]

17. Behm-Ansmant, I.; Kashima, I.; Rehwinkel, J.; Saulière, J.; Wittkopp, N.; Izaurralde, E. mRNA quality control: An ancient machinery recognizes and degrades mRNAs with nonsense codons. *Feb. Lett.* **2007**, *581*, 2845–2853. [CrossRef] [PubMed]

18. Chang, Y.-F.; Imam, J.S.; Wilkinson, M.F. The Nonsense-Mediated Decay RNA Surveillance Pathway. *Annu. Rev. Biochem.* **2007**, *76*, 51–74. [CrossRef]

19. Brogna, S.; Wen, J. Nonsense-mediated mRNA decay (NMD) mechanisms. *Nat. Struct. Mol. Biol.* **2009**, *16*, 107–113. [CrossRef]

20. Amrani, N.; Sachs, M.S.; Jacobson, A. Early nonsense: mRNA decay solves a translational problem. *Nat. Rev. Mol. Cell Biol.* **2006**, *7*, 415. [CrossRef]

21. Rebbapragada, I.; Lykke-Andersen, J. Execution of nonsense-mediated mRNA decay: what defines a substrate? *Curr. Opin. Cell Biol.* **2009**, *21*, 394–402. [CrossRef] [PubMed]

22. Peccarelli, M.; Kebaara, B.W. Regulation of Natural mRNAs by the Nonsense-Mediated mRNA Decay Pathway. *Eukaryot Cell* **2014**, *13*, 1126–1135. [CrossRef] [PubMed]

23. Kurihara, Y.; Matsui, A.; Hanada, K.; Kawashima, M.; Ishida, J.; Morosawa, T.; Tanaka, M.; Kaminuma, E.; Mochizuki, Y.; Matsushima, A.; et al. Genome-wide suppression of aberrant mRNA-like noncoding RNAs by NMD in Arabidopsis. *Proc. Natl. Acad. Sci. USA* **2009**, *106*, 2453–2458. [CrossRef] [PubMed]

24. Drechsel, G.; Kahles, A.; Kesarwani, A.K.; Stauffer, E.; Behr, J.; Drewe, P.; Rätsch, G.; Wachter, A. Nonsense-Mediated Decay of Alternative Precursor mRNA Splicing Variants Is a Major Determinant of the Arabidopsis Steady State Transcriptome. *Plant Cell* **2013**, *25*, 3726–3742. [CrossRef]

25. Kalyna, M.; Simpson, C.G.; Syed, N.H.; Lewandowska, D.; Marquez, Y.; Kusenda, B.; Marshall, J.; Fuller, J.; Cardle, L.; McNicol, J.; et al. Alternative splicing and nonsense-mediated decay modulate expression of important regulatory genes in Arabidopsis. *Nucleic Acids Res.* **2012**, *40*, 2454–2469. [CrossRef]

26. Leeds, P.; Wood, J.M.; Lee, B.S.; Culbertson, M.R. Gene products that promote mRNA turnover in Saccharomyces cerevisiae. *Mol. Cell Biol.* **1992**, *12*, 2165–2177. [CrossRef]

27. Kerényi, Z.; Mérai, Z.; Hiripi, L.; Benkovics, A.; Gyula, P.; Lacomme, C.; Barta, E.; Nagy, F.; Silhavy, D. Inter-kingdom conservation of mechanism of nonsense-mediated mRNA decay. *Embo J.* **2008**, *27*, 1585–1595. [CrossRef]

28. Shaul, O. Unique Aspects of Plant Nonsense-Mediated mRNA Decay. *Trends Plant Sci.* **2015**, *20*, 767–779. [CrossRef]

29. Rayson, S.; Arciga-Reyes, L.; Wootton, L.; De Torres Zabala, M.; Truman, W.; Graham, N.; Grant, M.; Davies, B. A role for nonsense-mediated mRNA decay in plants: Pathogen responses are induced in Arabidopsis thaliana NMD mutants. *PLoS ONE* **2012**, *7*, e31917. [CrossRef]

30. Shi, C.; Baldwin, I.T.; Wu, J. Arabidopsis Plants Having Defects in Nonsense-mediated mRNA Decay Factors UPF1, UPF2, and UPF3 Show Photoperiod-dependent Phenotypes in Development and Stress Responses. *J. Integr. Plant Biol.* **2012**, *54*, 99–114. [CrossRef]

31. Nasim, Z.; Fahim, M.; Ahn, J.H. Possible Role of MADS AFFECTING FLOWERING 3 and B-BOX DOMAIN PROTEIN 19 in Flowering Time Regulation of Arabidopsis Mutants with Defects in Nonsense-Mediated mRNA Decay. *Front. Plant Sci.* **2017**, *8*. [CrossRef]

32. Degtiar, E.; Fridman, A.; Gottlieb, D.; Berezin, I.; Vexler, K.; Golani, L.; Farhi, R.; Shaul, O. The feedback control of UPF3 is crucial for RNA surveillance in plants. *Nucleic Acids Res.* **2015**, *43*, 4219–4235. [CrossRef] [PubMed]

33. Popp, M.W.-L.; Maquat, L.E. Organizing Principles of Mammalian Nonsense-Mediated mRNA Decay. *Annu. Rev. Genet.* **2013**, *47*, 139–165. [CrossRef]

34. Dai, Y.; Li, W.; An, L. NMD mechanism and the functions of Upf proteins in plant. *Plant Cell Rep.* **2016**, *35*, 5–15. [CrossRef] [PubMed]

35. Karousis, E.D.; Mühlemann, O. Nonsense-Mediated mRNA Decay Begins Where Translation Ends. *Cold Spring Harb. Perspect. Biol.* **2019**, *11*. [CrossRef] [PubMed]

36. Doma, M.K.; Parker, R. Endonucleolytic cleavage of eukaryotic mRNAs with stalls in translation elongation. *Nature* **2006**, *440*, 561. [CrossRef] [PubMed]

37. Atkinson, G.C.; Baldauf, S.L.; Hauryliuk, V. Evolution of nonstop, no-go and nonsense-mediated mRNA decay and their termination factor-derived components. *BMC Evol. Biol.* **2008**, *8*, 290. [CrossRef] [PubMed]

38. Szádeczky-Kardoss, I.; Gál, L.; Auber, A.; Taller, J.; Silhavy, D. The No-go decay system degrades plant mRNAs that contain a long A-stretch in the coding region. *Plant Sci.* **2018**, *275*, 19–27. [CrossRef] [PubMed]

39. Shoemaker, C.J.; Eyler, D.E.; Green, R. Dom34:Hbs1 Promotes Subunit Dissociation and Peptidyl-tRNA Drop-Off to Initiate No-Go Decay. *Science* **2010**, *330*, 369–372. [CrossRef]

40. Tsuboi, T.; Kuroha, K.; Kudo, K.; Makino, S.; Inoue, E.; Kashima, I.; Inada, T. Dom34:Hbs1 Plays a General Role in Quality-Control Systems by Dissociation of a Stalled Ribosome at the 3′ End of Aberrant mRNA. *Mol. Cell* **2012**, *46*, 518–529. [CrossRef]

41. Beelman, C.A.; Parker, R. Differential effects of translational inhibition in cis and in trans on the decay of the unstable yeast MFA2 mRNA. *J. Biol. Chem.* **1994**, *269*, 9687–9692.

42. Buchan, J.R.; Stansfield, I. Halting a cellular production line: responses to ribosomal pausing during translation. *Biol. Cell* **2007**, *99*, 475–487. [CrossRef]

43. Simms, C.L.; Yan, L.L.; Zaher, H.S. Ribosome Collision Is Critical for Quality Control during No-Go Decay. *Mol. Cell* **2017**, *68*, 361–373. [CrossRef]

44. Ozsolak, F.; Kapranov, P.; Foissac, S.; Kim, S.W.; Fishilevich, E.; Monaghan, A.P.; John, B.; Milos, P.M. Comprehensive Polyadenylation Site Maps in Yeast and Human Reveal Pervasive Alternative Polyadenylation. *Cell* **2010**, *143*, 1018–1029. [CrossRef] [PubMed]

45. Dimitrova, L.N.; Kuroha, K.; Tatematsu, T.; Inada, T. Nascent Peptide-dependent Translation Arrest Leads to Not4p-mediated Protein Degradation by the Proteasome. *J. Biol. Chem.* **2009**, *284*, 10343–10352. [CrossRef]

46. Koutmou, K.S.; Schuller, A.P.; Brunelle, J.L.; Radhakrishnan, A.; Djuranovic, S.; Green, R. Ribosomes slide on lysine-encoding homopolymeric A stretches. *eLife* **2015**, *4*, e05534. [CrossRef]

47. van Hoof, A.; Frischmeyer, P.A.; Dietz, H.C.; Parker, R. Exosome-Mediated Recognition and Degradation of mRNAs Lacking a Termination Codon. *Science* **2002**, *295*, 2262–2264. [CrossRef] [PubMed]

48. Frischmeyer, P.A.; van Hoof, A.; O'Donnell, K.; Guerrerio, A.L.; Parker, R.; Dietz, H.C. An mRNA Surveillance Mechanism That Eliminates Transcripts Lacking Termination Codons. *Science* **2002**, *295*, 2258–2261. [CrossRef] [PubMed]

49. Auber, A.; Szádeczky-Kardoss, I.; Burgyán, J.; Csorba, T.; Nyikó, T.; Silhavy, D.; Schamberger, A.; Orbán, T.I.; Taller, J. The nonstop decay and the RNA silencing systems operate cooperatively in plants. *Nucleic Acids Res.* **2018**, *46*, 4632–4648. [CrossRef]

50. Oshima, T. A new polyamine, thermospermine, 1,12-diamino-4,8-diazadodecane, from an extreme thermophile. *J. Biol. Chem.* **1979**, *254*, 8720–8722. [PubMed]

51. Hanzawa, Y.; Takahashi, T.; Komeda, Y. ACL5: an Arabidopsis gene required for internodal elongation after flowering. *Plant J.* **1997**, *12*, 863–874. [CrossRef]

52. Hanzawa, Y.; Takahashi, T.; Michael, A.J.; Burtin, D.; Long, D.; Pineiro, M.; Coupland, G.; Komeda, Y. ACAULIS5, an Arabidopsis gene required for stem elongation, encodes a spermine synthase. *Embo J.* **2000**, *19*, 4248–4256. [CrossRef]

53. Knott, J.M.; Romer, P.; Sumper, M. Putative spermine synthases from Thalassiosira pseudonana and Arabidopsis thaliana synthesize thermospermine rather than spermine. *Febs. Lett.* **2007**, *581*, 3081–3086. [CrossRef] [PubMed]

54. Minguet, E.G.; Vera-Sirera, F.; Marina, A.; Carbonell, J.; Blazquez, M.A. Evolutionary Diversification in Polyamine Biosynthesis. *Mol. Biol. Evol.* **2008**, *25*, 2119–2128. [CrossRef] [PubMed]

55. Milhinhos, A.; Prestele, J.; Bollhöner, B.; Matos, A.; Vera-Sirera, F.; Rambla, J.L.; Ljung, K.; Carbonell, J.; Blázquez, M.A.; Tuominen, H.; et al. Thermospermine levels are controlled by an auxin-dependent feedback loop mechanism in Populus xylem. *Plant J.* **2013**, *75*, 685–698. [CrossRef]

56. Baima, S.; Forte, V.; Possenti, M.; Peñalosa, A.; Leoni, G.; Salvi, S.; Felici, B.; Ruberti, I.; Morelli, G. Negative Feedback Regulation of Auxin Signaling by ATHB8/ACL5–BUD2 Transcription Module. *Mol. Plant* **2014**, *7*, 1006–1025. [CrossRef] [PubMed]

57. Kakehi, J.i.; Kuwashiro, Y.; Niitsu, M.; Takahashi, T. Thermospermine is Required for Stem Elongation in Arabidopsis thaliana. *Plant Cell Physiol.* **2008**, *49*, 1342–1349. [CrossRef] [PubMed]

58. Clay, N.K.; Nelson, T. Arabidopsis thickvein Mutation Affects Vein Thickness and Organ Vascularization, and Resides in a Provascular Cell-Specific Spermine Synthase Involved in Vein Definition and in Polar Auxin Transport. *Plant Physiol.* **2005**, *138*, 767–777. [CrossRef]

59. Muñiz, L.; Minguet, E.G.; Singh, S.K.; Pesquet, E.; Vera-Sirera, F.; Moreau-Courtois, C.L.; Carbonell, J.; Blázquez, M.A.; Tuominen, H. ACAULIS5 controls Arabidopsis xylem specification through the prevention of premature cell death. *Development* **2008**, *135*, 2573–2582. [CrossRef]

60. Imai, A.; Hanzawa, Y.; Komura, M.; Yamamoto, K.T.; Komeda, Y.; Takahashi, T. The dwarf phenotype of the Arabidopsis acl5 mutant is suppressed by a mutation in an upstream ORF of a bHLH gene. *Development* **2006**, *133*, 3575–3585. [CrossRef]

61. Imai, A.; Komura, M.; Kawano, E.; Kuwashiro, Y.; Takahashi, T. A semi-dominant mutation in the ribosomal protein L10 gene suppresses the dwarf phenotype of the *acl5* mutant in *Arabidopsis thaliana*. *Plant J.* **2008**, *56*, 881–890. [CrossRef] [PubMed]

62. Kakehi, J.-I.; Kawano, E.; Yoshimoto, K.; Cai, Q.; Imai, A.; Takahashi, T. Mutations in Ribosomal Proteins, RPL4 and RACK1, Suppress the Phenotype of a Thermospermine-Deficient Mutant of Arabidopsis thaliana. *PLoS ONE* **2015**, *10*, e0117309. [CrossRef] [PubMed]

63. Cai, Q.; Fukushima, H.; Yamamoto, M.; Ishii, N.; Sakamoto, T.; Kurata, T.; Motose, H.; Takahashi, T. The SAC51 Family Plays a Central Role in Thermospermine Responses in Arabidopsis. *Plant Cell Physiol.* **2016**, *57*, 1583–1592. [CrossRef] [PubMed]

64. Vera-Sirera, F.; De Rybel, B.; Úrbez, C.; Kouklas, E.; Pesquera, M.; Álvarez-Mahecha, J.C.; Minguet, E.G.; Tuominen, H.; Carbonell, J.; Borst, J.W.; et al. A bHLH-Based Feedback Loop Restricts Vascular Cell Proliferation in Plants. *Dev. Cell* **2015**, *35*, 432–443. [CrossRef] [PubMed]

65. Yamamoto, M.; Takahashi, T. Thermospermine enhances translation of SAC51 and SACL1 in Arabidopsis. *Plant Signal. Behav.* **2017**, *12*, e1276685. [CrossRef]

66. von Arnim, A.G.; Jia, Q.; Vaughn, J.N. Regulation of plant translation by upstream open reading frames. *Plant Sci.* **2014**, *214*, 1–12. [CrossRef]

67. Weiss, M.C.; Sousa, F.L.; Mrnjavac, N.; Neukirchen, S.; Roettger, M.; Nelson-Sathi, S.; Martin, W.F. The physiology and habitat of the last universal common ancestor. *Nat. Microbiol.* **2016**, *1*, 16116. [CrossRef]

68. Imai, A.; Matsuyama, T.; Hanzawa, Y.; Akiyama, T.; Tamaoki, M.; Saji, H.; Shirano, Y.; Kato, T.; Hayashi, H.; Shibata, D.; et al. Spermidine Synthase Genes Are Essential for Survival of Arabidopsis. *Plant Physiol.* **2004**, *135*, 1565–1573. [CrossRef]

69. Hamasaki-Katagiri, N.; Tabor, C.W.; Tabor, H. Spermidine biosynthesis in Saccharomyces cerevisiae: Polyamine requirement of a null mutant of the SPE3 gene (spermidine synthase). *Gene* **1997**, *187*, 35–43. [CrossRef]

70. Mandal, S.; Mandal, A.; Johansson, H.E.; Orjalo, A.V.; Park, M.H. Depletion of cellular polyamines, spermidine and spermine, causes a total arrest in translation and growth in mammalian cells. *Proc. Natl. Acad. Sci. USA* **2013**, *110*, 2169–2174. [CrossRef]

71. Park, M.H.; Wolff, E.C. Hypusine, a polyamine-derived amino acid critical for eukaryotic translation. *J. Biol. Chem.* **2018**. [CrossRef]

72. Park, M. The post-translational synthesis of a polyamine-derived amino acid, hypusine, in the eukaryotic translation initiation factor 5A (eIF5A). *J. Biochem.* **2006**, *139*, 161–169. [CrossRef] [PubMed]

73. Park, M.H.; Wolff, E.C.; Folk, J.E. Hypusine - Its Posttranslational Formation in Eukaryotic Initiation Factor-5A and Its Potential Role in Cellular-Regulation. *Biofactors* **1993**, *4*, 95–104. [PubMed]

74. Chattopadhyay, M.K.; Park, M.H.; Tabor, H. Hypusine modification for growth is the major function of spermidine in *Saccharomyces cerevisiae* polyamine auxotrophs grown in limiting spermidine. *Proc. Natl. Acad. Sci. USA* **2008**, *105*, 6554–6559. [CrossRef] [PubMed]

75. Pällmann, N.; Braig, M.; Sievert, H.; Preukschas, M.; Hermans-Borgmeyer, I.; Schweizer, M.; Nagel, C.H.; Neumann, M.; Wild, P.; Haralambieva, E.; et al. Biological Relevance and Therapeutic Potential of the Hypusine Modification System. *J. Biol. Chem.* **2015**, *290*, 18343–18360. [CrossRef] [PubMed]

76. Nishimura, K.; Lee, S.; Park, J.; Park, M. Essential role of eIF5A-1 and deoxyhypusine synthase in mouse embryonic development. *Amino Acids* **2012**, *42*, 703–710. [CrossRef] [PubMed]

77. Pagnussat, G.C.; Yu, H.J.; Ngo, Q.A.; Rajani, S.; Mayalagu, S.; Johnson, C.S.; Capron, A.; Xie, L.F.; Ye, D.; Sundaresan, V. Genetic and molecular identification of genes required for female gametophyte development and function in Arabidopsis. *Development* **2005**, *132*, 603–614. [CrossRef]

78. Thomas, A.; Goumans, H.; Amesz, H.; Benne, R.; Voorma, H.O. A Comparison of the Initiation Factors of Eukaryotic Protein Synthesis from Ribosomes and from the Postribosomal Supernatant. *Eur. J. Biochem.* **1979**, *98*, 329–337. [CrossRef]

79. Cooper, H.L.; Park, M.H.; Folk, J.E.; Safer, B.; Braverman, R. Identification of the hypusine-containing protein hy+ as translation initiation factor eIF-4D. *Proc. Natl. Acad. Sci. USA* **1983**, *80*, 1854–1857. [CrossRef]

80. Kemper, W.M.; Berry, K.W.; Merrick, W.C. Purification and properties of rabbit reticulocyte protein synthesis initiation factors M2Balpha and M2Bbeta. *J. Biol. Chem.* **1976**, *251*, 5551–5557.

81. Shiba, T.; Mizote, H.; Kaneko, T.; Nakajima, T.; Yasuo, K.; Sano, I. Hypusine, a new amino acid occurring in bovine brain: Isolation and structural determination. *Biochim. Biophys. Acta (BBA) Gen. Subj.* **1971**, *244*, 523–531. [CrossRef]

82. Park, M.H.; Cooper, H.L.; Folk, J.E. Identification of hypusine, an unusual amino acid, in a protein from human lymphocytes and of spermidine as its biosynthetic precursor. *Proc. Natl. Acad. Sci. USA* **1981**, *78*, 2869–2873. [CrossRef]

83. Saini, P.; Eyler, D.E.; Green, R.; Dever, T.E. Hypusine-containing protein eIF5A promotes translation elongation. *Nature* **2009**, *459*, 118–121. [CrossRef]

84. Schuller, A.P.; Wu, C.C.-C.; Dever, T.E.; Buskirk, A.R.; Green, R. eIF5A Functions Globally in Translation Elongation and Termination. *Mol. Cell* **2017**, *66*, 194–205.e195. [CrossRef] [PubMed]

85. Gäbel, K.; Schmitt, J.; Schulz, S.; Näther, D.J.; Soppa, J. A Comprehensive Analysis of the Importance of Translation Initiation Factors for Haloferax volcanii Applying Deletion and Conditional Depletion Mutants. *PLoS ONE* **2013**, *8*, e77188. [CrossRef] [PubMed]

86. Kyrpides, N.C.; Woese, C.R. Universally conserved translation initiation factors. *Proc. Natl. Acad. Sci. USA* **1998**, *95*, 224–228. [CrossRef]

87. Navarre, W.W.; Zou, S.B.; Roy, H.; Xie, J.L.; Savchenko, A.; Singer, A.; Edvokimova, E.; Prost, L.R.; Kumar, R.; Ibba, M.; et al. PoxA, YjeK, and Elongation Factor P Coordinately Modulate Virulence and Drug Resistance in *Salmonella enterica. Mol. Cell* **2010**, *39*, 209–221. [CrossRef]

88. Lassak, J.; Keilhauer, E.C.; Fürst, M.; Wuichet, K.; Gödeke, J.; Starosta, A.L.; Chen, J.-M.; Søgaard-Andersen, L.; Rohr, J.; Wilson, D.N.; et al. Arginine-rhamnosylation as new strategy to activate translation elongation factor P. *Nat. Chem. Biol.* **2015**, *11*, 266–270. [CrossRef]

89. Bullwinkle, T.J.; Zou, S.B.; Rajkovic, A.; Hersch, S.J.; Elgamal, S.; Robinson, N.; Smil, D.; Bolshan, Y.; Navarre, W.W.; Ibba, M. (R)-β-Lysine-modified Elongation Factor P Functions in Translation Elongation. *J. Biol. Chem.* **2013**, *288*, 4416–4423. [CrossRef]

90. Balibar, C.J.; Iwanowicz, D.; Dean, C.R. Elongation Factor P is Dispensable in Escherichia coli and Pseudomonas aeruginosa. *Curr. Microbiol.* **2013**, *67*, 293–299. [CrossRef]

91. Blaha, G.; Stanley, R.E.; Steitz, T.A. Formation of the First Peptide Bond: The Structure of EF-P Bound to the 70S Ribosome. *Science* **2009**, *325*, 966–970. [CrossRef]

92. Melnikov, S.; Mailliot, J.; Shin, B.-S.; Rigger, L.; Yusupova, G.; Micura, R.; Dever, T.E.; Yusupov, M. Crystal Structure of Hypusine-Containing Translation Factor eIF5A Bound to a Rotated Eukaryotic Ribosome. *J. Mol. Biol.* **2016**. [CrossRef] [PubMed]

93. Schmidt, C.; Becker, T.; Heuer, A.; Braunger, K.; Shanmuganathan, V.; Pech, M.; Berninghausen, O.; Wilson, D.N.; Beckmann, R. Structure of the hypusinylated eukaryotic translation factor eIF-5A bound to the ribosome. *Nucleic Acids Res.* **2016**, *44*, 1944–1951. [CrossRef] [PubMed]

94. Gutierrez, E.; Shin, B.-S.; Woolstenhulme, C.J.; Kim, J.-R.; Saini, P.; Buskirk, A.R.; Dever, T.E. eIF5A Promotes Translation of Polyproline Motifs. *Mol. Cell* **2013**, *51*, 35–45. [CrossRef]

95. Doerfel, L.K.; Wohlgemuth, I.; Kothe, C.; Peske, F.; Urlaub, H.; Rodnina, M.V. EF-P Is Essential for Rapid Synthesis of Proteins Containing Consecutive Proline Residues. *Science* **2013**, *339*, 85–88. [CrossRef] [PubMed]

96. Ude, S.; Lassak, J.; Starosta, A.L.; Kraxenberger, T.; Wilson, D.N.; Jung, K. Translation Elongation Factor EF-P Alleviates Ribosome Stalling at Polyproline Stretches. *Science* **2013**, *339*, 82–85. [CrossRef] [PubMed]

97. Pavlov, M.Y.; Watts, R.E.; Tan, Z.; Cornish, V.W.; Ehrenberg, M.; Forster, A.C. Slow peptide bond formation by proline and other N-alkylamino acids in translation. *Proc. Natl. Acad. Sci. USA* **2009**, *106*, 50–54. [CrossRef] [PubMed]

98. Belda Palazón, B.; Almendáriz, C.; Martí, E.; Carbonell, J.; Ferrando, A. Relevance of the axis spermidine/eIF5A for plant growth and development. *Front. Plant Sci.* **2016**, *7*. [CrossRef] [PubMed]

99. Li, T.; Belda-Palazon, B.; Ferrando, A.; Alepuz, P. Fertility and Polarized Cell Growth Depends on eIF5A for Translation of Polyproline-Rich Formins in *Saccharomyces cerevisiae*. *Genetics* **2014**, *197*, 1191–1200. [CrossRef]

100. Duguay, J.; Jamal, S.; Liu, Z.; Wang, T.W.; Thompson, J.E. Leaf-specific suppression of deoxyhypusine synthase in *Arabidopsis thaliana* enhances growth without negative pleiotropic effects. *J. Plant Physiol.* **2007**, *164*, 408–420. [CrossRef]

101. Feng, H.; Chen, Q.; Feng, J.; Zhang, J.; Yang, X.; Zuo, J. Functional Characterization of the Arabidopsis Eukaryotic Translation Initiation Factor 5A-2 That Plays a Crucial Role in Plant Growth and Development by Regulating Cell Division, Cell Growth, and Cell Death. *Plant Physiol.* **2007**, *144*, 1531–1545. [CrossRef] [PubMed]

102. Liu, Z.; Duguay, J.; Ma, F.; Wang, T.W.; Tshin, R.; Hopkins, M.T.; McNamara, L.; Thompson, J.E. Modulation of eIF5A1 expression alters xylem abundance in *Arabidopsis thaliana*. *J. Exp. Bot.* **2008**, *59*, 939–950. [CrossRef] [PubMed]

103. Ma, F.; Liu, Z.; Wang, T.W.; Hopkins, M.T.; Peterson, C.A.; Thompson, J.E. Arabidopsis eIF5A3 influences growth and the response to osmotic and nutrient stress. *Plant Cell Env.* **2010**, *33*, 1682–1696. [CrossRef] [PubMed]

104. Buskirk, A.R.; Green, R. Ribosome pausing, arrest and rescue in bacteria and eukaryotes. *Philos. Trans. R. Soc. B: Biol. Sci.* **2017**, *372*. [CrossRef]

105. Dever, T.E.; Dinman, J.D.; Green, R. Translation Elongation and Recoding in Eukaryotes. *Cold Spring Harb. Perspect. Biol.* **2018**, *10*. [CrossRef] [PubMed]

106. Zuk, D.; Jacobson, A. A single amino acid substitution in yeast eIF-5A results in mRNA stabilization. *Embo J.* **1998**, *17*, 2914–2925. [CrossRef]

107. Schrader, R.; Young, C.; Kozian, D.; Hoffmann, R.; Lottspeich, F. Temperature-sensitive eIF5A Mutant Accumulates Transcripts Targeted to the Nonsense-mediated Decay Pathway. *J. Biol. Chem.* **2006**, *281*, 35336–35346. [CrossRef]

108. Hoque, M.; Park, J.Y.; Chang, Y.-J.; Luchessi, A.D.; Cambiaghi, T.D.; Shamanna, R.; Hanauske-Abel, H.M.; Holland, B.; Pe'ery, T.; Tian, B.; et al. Regulation of gene expression by translation factor eIF5A: Hypusine-modified eIF5A enhances nonsense-mediated mRNA decay in human cells. *Translation* **2017**, *5*, e1366294. [CrossRef]

109. Li, C.H.; Ohn, T.; Ivanov, P.; Tisdale, S.; Anderson, P. eIF5A Promotes Translation Elongation, Polysome Disassembly and Stress Granule Assembly. *PLoS ONE* **2010**, *5*, e9942. [CrossRef]

110. Miller-Fleming, L.; Olin-Sandoval, V.; Campbell, K.; Ralser, M. Remaining Mysteries of Molecular Biology: The Role of Polyamines in the Cell. *J. Mol. Biol.* **2015**, *427*, 3389–3406. [CrossRef] [PubMed]

Review

Polyamine Oxidases Play Various Roles in Plant Development and Abiotic Stress Tolerance

Zhen Yu [1], Dongyu Jia [2] and Taibo Liu [1,*]

[1] State Key Laboratory for Conservation and Utilization of Subtropical Agro-Bioresources, Guangdong Provincial Key Laboratory of Protein Function and Regulation in Agricultural Organisms, College of Life Sciences, South China Agricultural University, Guangzhou 510642, China; yuzhen5500@163.com

[2] Department of Biology, Georgia Southern University, Statesboro, GA 30460-8042, USA; djia@georgiasouthern.edu

[*] Correspondence: tbliu@scau.edu.cn; Tel.: +86-20-3829-7785

Received: 29 April 2019; Accepted: 20 June 2019; Published: 21 June 2019

Abstract: Polyamines not only play roles in plant growth and development, but also adapt to environmental stresses. Polyamines can be oxidized by copper-containing diamine oxidases (CuAOs) and flavin-containing polyamine oxidases (PAOs). Two types of PAOs exist in the plant kingdom; one type catalyzes the back conversion (BC-type) pathway and the other catalyzes the terminal catabolism (TC-type) pathway. The catabolic features and biological functions of plant PAOs have been investigated in various plants in the past years. In this review, we focus on the advance of PAO studies in rice, Arabidopsis, and tomato, and other plant species.

Keywords: back conversion pathway; polyamines; polyamine oxidase; polyamine catabolism; stress response; terminal catabolism pathway

---

1. Introduction

Polyamines (PAs) are aliphatic amines of small molecular mass that are involved in various biological processes [1,2]. The putrescine (Put), cadaverine (Cad), spermidine (Spd), spermine (Spm), and thermospermine (T-Spm) are the major plant PAs [1–7]. PAs play important roles in embryogenesis, cell division, organogenesis, flowering, programmed cell death (PCD), response to abiotic and biotic stresses, and so on [4–33].

The homeostasis of cellular PA levels, being well regulated by a dynamic balance of biosynthesis and catabolism, is most important for maintaining normal growth and development in plants. The PA biosynthetic pathway has been well elucidated [1,34,35], however, the PA catabolism pathway remains unclear in spite of more and more newly identified genes in this pathway in plants [4,36–56]. In this review, we summarized the advances of the polyamine oxidases' (PAOs) roles in PA catabolism, plant development, and abiotic stress tolerance from rice, Arabidopsis, tomato, and other plant species.

2. PA Biosynthesis in Plants

Plant PA biosynthesis is rather short, which starts mainly from arginine (Arg). The pathway is briefly shown in Figure 1 and is described as follows. Firstly, Arg is converted to Put via agmatine by three sequential reactions catalyzed by arginine decarboxylase (ADC, EC 4.1.1.19), agmatine iminohydrolase (AIH, EC 3.5.3.12), and *N*-carbamoylputrescine amidohydrolase (CPA, EC 3.5.1.53). Besides, some plants have the ornithine decarboxylase (ODC, EC 4.1.1.17) which catalyzes ornithine to Put directly [57], but Arabidopsis has only the ADC pathway because it lacks *ODC* genes. Secondly, the diamine Put is converted to triamine Spd by Spd synthase (SPDS, EC 2.5.1.16). Finally, Spd is further converted to Spm or T-Spm, two tetraamine isomers, by Spm synthase (SPMS, EC 2.5.1.22) and T-Spm synthase (ACAULIS5, abbreviated to ACL5), respectively [9,19,47,58]. An aminopropyl group is transferred from the decarboxylated *S*-adenosylmethionine (dcSAM) produced from methionine in two sequential reactions catalyzed by methionine adenosyltransferase and *S*-adenosylmethionine decarboxylase (SAMDC), respectively. These aminopropyl groups participate in the biochemical reaction of Spd, Spm, and T-Spm biosynthesis processes. Additionally, norspermidine (NorSpd) and norspermine (NorSpm), having been found as "uncommon PAs" due to their limited distribution in nature, are predicted to be synthesized either successively by each specific aminopropyl transferase (APT) or by a single APT with broad substrate specificity from 1,3-diaminopropane (1,3-DAP) [59].

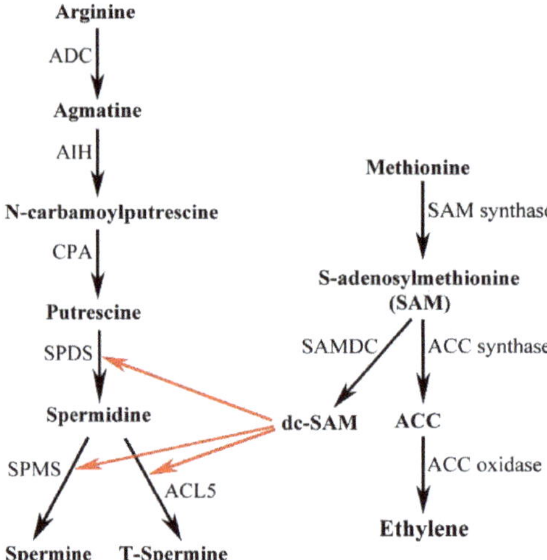

Figure 1. Polyamine biosynthesis pathway in *Arabidopsis thaliana*. ADC, arginine decarboxylase; AIH, agmatine iminohydrolase; CPA, *N*-carbamoylputrescine amidohydrolase; SPDS, Spd synthase; SPMS, Spm synthase; ACL5, ACAULIS5, T-Spm synthase; SAM, *S*-adenosylmethionine; SAMDC, *S*-adenosylmethionine decarboxylase; dcSAM, decarboxylated *S*-adenosylmethionine; ACC, 1-amino-cyclopropane-1-carboxylic-acid.

3. PA Catabolism in Plants

PA biosynthetic pathways have been well investigated. In contrast, the knowledge on PA catabolism in plants is still fragmental though scholars reported some new findings in the past years. Two kinds of enzymes are involved in PA catabolism. Namely, one is a copper-dependent diamine oxidase (DAO, EC 1.4.3.6) and the other is a flavin adenine dinucleotide (FAD)-dependent polyamine oxidase (PAO, EC 1.5.3.11). PAOs, using FAD as cofactor, catalyze Spd and Spm to produce 4-aminobutanal and *N*-(3-aminopropyl)-4-aminobutanal, respectively, as well as hydrogen peroxide (H_2O_2) which acts as an important signaling to regulate the expression of numerous genes relative to the stress response in the back conversion (BC-type) pathway; in addition to 1,3-diaminopropane and H_2O_2 in the terminal catabolism (TC-type) pathway [46–49,51–54,60].

4. PAOs in Plants

Up to now, more and more plant PAOs have been cloned and functionally identified. In Figure 2, we analyzed the phylogenetic relationship among seventy-three plant PAOs from twenty-four species. The plant PAOs are grouped into five clades I~V in the phylogenetic tree, as shown in Figure 2. Clade-I has nine members including Arabidopsis PAO (AtPAO1) and tomato PAO (SlPAO1) [48,61–63]. Clade-II contains sixteen genes including three rice PAOs (OsPAO2, OsPAO6~7) [48,60,63]. Based on previous studies, the clade II may present apoplastic PAOs that catalyze terminal oxidation reactions [36,42,44–46,49,55,60]. Clade-III consists of nineteen members including rice PAO (OsPAO1), Arabidopsis PAO

(AtPAO5), and two tomato PAOs (SlPAO6~7) [49,51,63]. Clade-IV contains twenty-eight PAOs from eight different species including three rice PAOs (OsPAO3~5), three Arabidopsis PAOs (AtPAO2~4), and four tomato PAOs (SlPAO2~5) [12,47,48,50,61,63,64]. The clade V so far includes only a *Vitis vinifera* PAO (VvPAO6). Currently, almost all PAOs of the rice and Arabidopsis have been well determined, and we recently identified the tomato PAOs. Thus, we will review on the advance of PAOs from these three species, as well as other plant species, in this manuscript.

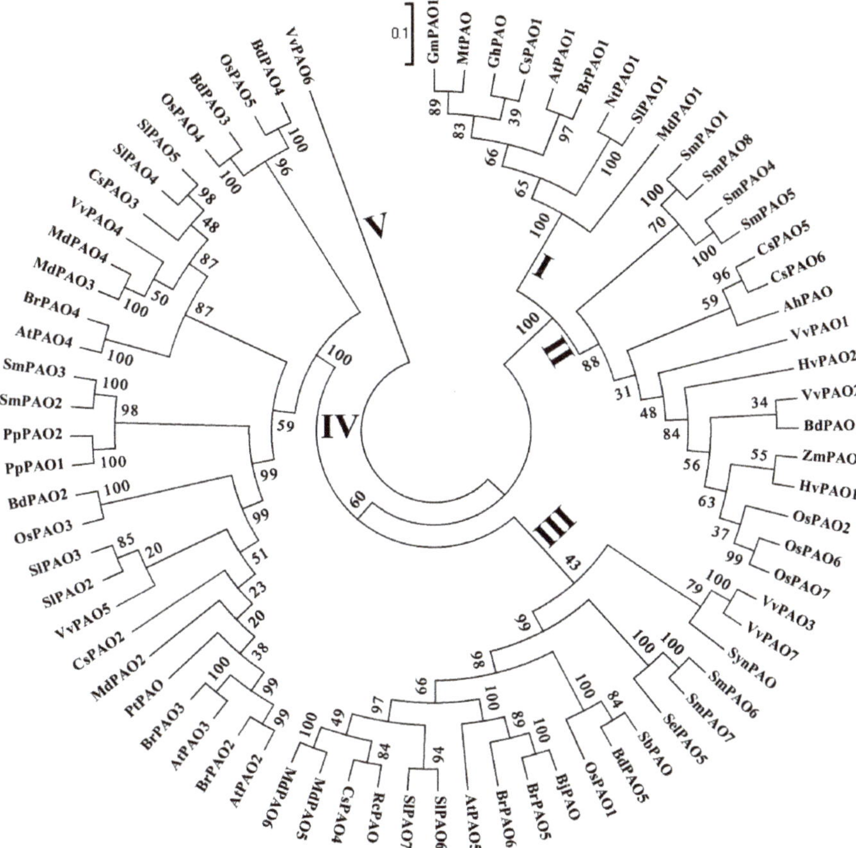

Figure 2. Phylogenetic relationship of polyamine oxidases (PAOs) among rice, Arabidopsis, tomato, and other plants. The neighbor-joining tree was constructed by amino acid sequence alignment using Clustal X 1.83 and MEGA 5.0. The bootstrap values, displayed at the branch nodes, were obtained with 1000 repetitions. Roman numerals (I~V) indicate clade numbers. The analyzed genes and their accession numbers are listed in Table 1. Os, *Oryza sativa*; At, *Arabidopsis thaliana*; Sl, *Solanum lycopersicum*; Bd, *Brachypodium distachyon*; Br, *Solanum lycopersicum*; Cs, *Citrus sinensis*; Sm, *Selaginella moellendorffii*; Vv, *Vitis vinifera*; Md, *Malus domestica*; Sel, *Selaginella lepidophylla*; Zm, *Zea mays*; Hv, *Hordeum vulgare*; Pp, *Physcomitrella patens*; Rc, *Ricinus communis*; Nt, *Nicotiana tabacum*; Bj, *Brassica juncea*; Pt, *Populus trichocarpa*; Sb, *Sorghum bicolor*; Gm, *Glycine max PAO1-like*; Mt, *Medicago truncatula*; Ah, *Amaranthus hypochondriacus*; Gh, *Gossypium hirsutum*; Syn, *Synechocystis*.

Table 1. List of the accession numbers of the plant PAOs used in Figure 2.

Gene Name	Accession No.	Gene Name	Accession No.	Gene Name	Accession No.	Gene Name	Accession No.
OsPAO1	NM_001050573	*BdPAO1*	XM_003573843	*SmPAO3*	XP_002968082.1	*PpPAO2*	XM_001776435
OsPAO2	NM_001055782	*BdPAO2*	XM_010242147	*SmPAO4*	XP_002969966.1	*RcPAO*	XM_002521542
OsPAO3	NM_001060458	*BdPAO3*	XM_003580746	*SmPAO5*	XP_002981437.1	*PtPAO*	XM_002306729
OsPAO4	NM_001060753	*BdPAO4*	XM_003580747	*SmPAO6*	XP_002984796.1	*SbPAO*	XM_002448510
OsPAO5	NM_001060754	*BdPAO5*	XM_003566997	*SmPAO7*	XP_002985859.1	*GmPAO1*	XP_003535841.1
OsPAO6	XM_015755533	*BrPAO1*	Bra006210	*SmPAO8*	XP_002986593.1	*MtPAO*	XP_003599417.1
OsPAO7	NM_001069546	*BrPAO2*	Bra037741	*VvPAO1*	VIT_01s0127g00750	*SynPAO*	WP_011153630.1
AtPAO1	NM_121373	*BrPAO3*	Bra003362	*VvPAO2*	VIT_01s0127g00800	*PpPAO1*	XM_001756812
AtPAO2	AF364952	*BrPAO4*	Bra039742	*VvPAO3*	VIT_03s0017g01000	*ZmPAO1*	NM_001111636
AtPAO3	AY143905	*BrPAO5*	Bra011132	*VvPAO4*	VIT_04s0043g00220	*AhPAO*	AAM43922.1
AtPAO4	AF364953	*BrPAO6*	Bra024137	*VvPAO5*	VIT_12s0028g01120	*GhPAO*	KC762210.1
AtPAO5	AK118203	*CsPAO1*	Cs7g02060.1	*VvPAO6*	VIT_12s0055g00480	*HvPAO1*	AJ298131
SlPAO1	XP_004229651	*CsPAO2*	Cs7g18840.2	*VvPAO7*	VIT_13s0019g04820	*HvPAO2*	AJ298132
SlPAO2	XP_004243630	*CsPAO3*	Cs6g15870.1	*MdPAO1*	ANJ77637.1	*SelPAO5*	LC036642
SlPAO3	XP_004251556	*CsPAO4*	Cs4g14150.1	*MdPAO2*	ANJ77639.1	*NtPAO*	AB200262
SlPAO4	XP_004232664	*CsPAO5*	Cs7g23790.1	*MdPAO3*	ANJ77642.1	*BjPAO*	AY188087
SlPAO5	XP_004234492	*CsPAO6*	Cs7g23760.1	*MdPAO4*	ANJ77638.1		
SlPAO6	XP_004243758	*SmPAO1*	XP_002965265.1	*MdPAO5*	ANJ77640.1		
SlPAO7	XP_004239292	*SmPAO2*	XP_002965599.1	*MdPAO6*	ANJ77641.1		

4.1. Rice PAOs

Ono et al. reported that seven *PAOs* exist in rice, orderly named as *OsPAO1~OsPAO7* [47]. He and his colleagues found *OsPAO3~5* are similarly and highly expressed in two-week-old seedlings and mature plants, whereas the other four *OsPAO* members are only expressed at very low levels in all tissues. Especially, *OsPAO2*, *OsPAO6*, and *OsPAO7* are expressed at almost negligible levels, as shown in Table 2 [47,49]. They also found the purified recombinant OsPAO3 strongly catalyzes Spd to Put, and also utilizes Spm, T-Spm, and Nor-Spm as substrates in vivo. The OsPAO4 and OsPAO5 proteins prefer to use Spm and T-Spm as substrates, but cannot oxidize Spd to Put, as shown in Table 2 [46,47]. The results suggested that OsPAO3 catalyzes a full BC-type pathway, while OsPAO4 and OsPAO5 only catalyze a partial BC-type pathway, as shown in Table 2 [46,47]. Besides, we found that OsPAO1, localized to the cytoplasm of onion epidermal cells, prefers to use Spm and T-Spm as substrates, and oxidizes these substrates to Spd but not to Put, as shown in Table 2 [46,48]. OsPAO1 and AtPAO5, both of which lack of intron, share high identity at the amino acid levels and exhibit quite similar predicted protein tertiary structures [50]. When the full length cDNA of OsPAO1 was fused to a constitutive promoter and subsequently transformed into the loss-of-function mutant *Atpao5-2*, the transgenic plants restored normal T-Spm sensitivity, which can grow in the presence of low levels of T-Spm; whereas the control with the introduction of *OsPAO3*—a peroxisome localized *PAO*—into *Atpao5-2* mutants did not complement the phenotype [50]. These genetic evidences indicated that *OsPAO1* and *AtPAO5* are functionally orthologous genes in Arabidopsis and rice [50].

Interestingly, our group found that OsPAO7, with high amino acid identity and very similarly predicted protein 3-D structures to ZmPAO1, which is the best characterized maize PAO catalyzed TC-type reaction, is subcellularly localized to the apoplastic space with the aid of a signal peptides (SPs, amino acid position 1-19) and transmembrane domains (TDs, amino acid position 20-29) in its N-terminal, as shown in Table 2 [46,49]. The recombinant OsPAO7 produces 1,3-diaminopropane from both Spd and Spm, indicating that OsPAO7 is the first TC-type enzyme in rice, as shown in Table 2 [46,49]. The observation of *OsPAO7$_{pro}$:GFP* transgenic rice plants showed that *OsPAO7* is specifically expressed in anther walls and pollens with an expressional peak at the bicellular pollen stages, as shown in Table 2 [46,49]. Such results suggest that *OsPAO7* might have special roles in floral differentiation, especially in anther development and fertility, as shown in Table 2. Recently, Sagor et al. reported that the DNA sequence of the presumed coding region (accession number NM_001069545) for *OsPAO6* obtained from the National Center for Biotechnology Information (NCBI) public database is incorrect [60]. They successfully cloned the correct full-length cDNA of 1742 bp (accession number XM_015755533) by rapid amplification of the cDNA ends (RACE) in the 5′-end using 5′-RACE [60]. The correct *OsPAO6*, encoding a 497-amino acid protein, shows 92% identity and very similar protein tertiary structures to *OsPAO7*, and it is subcellularly localized to the plasma membrane, suggesting that OsPAO6 possibly also acts like OsPAO7 having the TC-type activity [46,49,60]. Furthermore, *OsPAO6* was induced by exogenous jasmonic acid, implying *OsPAO6* may be involved in stress tolerance [60]. The last rice PAO, OsPAO2, might have no enzyme activity due to a long truncation at the amino terminal [46,49,60]. However, we could not rule out the possibility that the cDNA sequence of *OsPAO2* derived from NCBI might be incorrect like the case of OsPAO7.

Up to now, the knowledge of the biological functions of *OsPAOs* remains limited. Chen et al. found that *OsPAO1-7* is most important for rice germination compared to the subfamilies' members *OsPAO8~11* encoding histone lysine-specific demethylases, especially *OsPAO5* which probably regulates rice seed germination via PAO-generated H_2O_2 signaling to mediate coleorhiza-limited rice seed germination [65].

Table 2. Summary of PAOs in rice, Arabidopsis, and tomato.

Gene Name	Gene ID	Subcellular Localization	Substrate Specificity	Mode of Reaction	Tissue Expression	Functions (or Potential Functions)	Reference
					Oryza sativa		
OsPAO1	Os01g0710200	cytoplasm	Spm, T-Spm	BC	rachis	rachis development, tolerances, seed germination	[31,46–48]
OsPAO2	Os03g0193400	n.d.	n.d.	n.d.	root (with very low expression levels)	tolerances, seed germination	[31,46,49]
OsPAO3	Os04g0623300	peroxisome	Spd, Spm, T-Spm	BC	All stages. Strongest expressed in leaf, rachis, node, lower leaf blade, mature floral organ	leaf and node development, floral development, fertility, seed germination	[31,46,47]
OsPAO4	Os04g0671200	peroxisome	Spm, T-Spm	BC	rachis, mature floral organ	rachis and floral development, fertility, seed germination	[31,46,47]
OsPAO5	Os04g0671300	peroxisome	Spm, T-Spm	BC	flag leaf, lower leaf blade, leaf sheath, mature floral organ	development of leaf and flower, fertility, seed germination	[31,46,47]
OsPAO6	Os09g0368200	apoplast	n.d.	TC (?)	expressed at negligible levels	tolerances, seed germination	[31,46,60]
OsPAO7	Os09g0368500	apoplast	Spm, Spd	TC	anther, pollen	floral development, fertility, seed germination	[31,46,49]
					Arabidopsis thaliana		
AtPAO1	At5g13700	cytoplasm	Spm, T-Spm	BC	root transition region, anther	stress tolerance, root development, fertility	[39,46,61,62,65]
AtPAO2	At2g43020	peroxisome	Spd, Spm, T-Spm	BC	root meristem, anther, main vein of rosette leaf	root development, fertility, vein development of leaf	[46,61,62,64,65]
AtPAO3	At3g59050	peroxisome	Spd, Spm, T-Spm	BC	All stages. Strongest expressed in root tip, flower, guard cell	root and leaf development, fertility	[12,46,61,62,65]
AtPAO4	At1g65840	peroxisome	Spm, T-Spm	BC	All stages. Strongest expressed in root and floral organ	Delay dark-induced senescence. Root development, fertility	[46,61,62,64–66]
AtPAO5	At4g29720	cytoplasm	Spm, T-Spm	BC	All stages. Strongest expressed in mature leaf, vascular tissue, flower, stem	xylem differentiation, stem elongation, development of rosette leaves and vein, tolerance	[46,51,61,62,65,67–69]

Table 2. *Cont.*

Gene Name	Gene ID	Subcellular Localization	Substrate Specificity	Mode of Reaction	Tissue Expression	Functions (or Potential Functions)	Reference
					Solanum lycopersicum		
SlPAO1	Solyc01g087590	n.d.	n.d.	n.d.	root, stem, leaf of seedling stage	vegetative growth	[63]
SlPAO2	Solyc07g043590	peroxisome (?)	n.d.	n.d.	All stages. Strongest expressed in anther, Br, Br+2, stem	floral development, fruit maturity	[63]
SlPAO3	Solyc12g006370	peroxisome (?)	n.d.	n.d.	All stages. Strongest expressed in anther, Br, Br+2, leaf	floral development, fruit maturity	[63]
SlPAO4	Solyc02g081390	peroxisome (?)	n.d.	n.d.	All stages. Strongest expressed in anther, Br, Br+2, Br+7, root, leaf	floral development, fruit maturity	[63]
SlPAO5	Solyc03g031880	peroxisome (?)	n.d.	n.d.	All stages. Strongest expressed in anther, leaf, stem	floral development	[63]
SlPAO6	Solyc07g039310	n.d.	n.d.	n.d.	root, stem of seedling stage	vegetative growth	[63]
SlPAO7	Solyc05g018880	n.d.	n.d.	n.d.	root, stem of seedling stage	vegetative growth	[63]
					Brachypodium distachyon		
BdPAO1	XM_003573843	n.d.	n.d.	n.d.	expressed at very low levels	unknown	[70]
BdPAO2	XM_010242147	peroxisome (?)	Spd, Spm, T-Spm, Nor-Spm, Nor-Spd	BC	All stages. Highly expressed in leaf, stem, and inflorescence	development of stem and inflorescence	[70]
BdPAO3	XM_003580746	n.d.	Spm,	BC	leaf, stem, and inflorescence	development of stem and inflorescence	[70]
BdPAO4	XM_003580747	peroxisome (?)	n.d.	n.d.	leaf, stem, and inflorescence	development of stem and inflorescence	[70]
BdPAO5	XM_003566997	n.d.	n.d.	n.d.	expressed at very low levels	unknown	[70]
					Citrus sinensis		
CsPAO1	Cs7g02060.1	n.d.	n.d.	BC (?)	leaf, stem, root, cotyledon	root growth, vegetative growth	[55,71]
CsPAO2	Cs7g18840.2	peroxisome (?)	n.d.	BC (?)	leaf, stem, root, cotyledon	root growth, vegetative growth	[55,71]
CsPAO3	Cs6g15870.1	peroxisome (?)	n.d.	BC (?)	leaf, stem, root, cotyledon	root growth, vegetative growth	[55,71]
CsPAO4	Cs4g14150.1	apoplast	Spd, Spm	TC	leaf, stem, root	seed germination, the growth of root and vegetative, salt tolerance	[55,71]
CsPAO5	Cs7g23790.1	n.d.	n.d.	BC (?)	leaf, stem, root, cotyledon	root growth, vegetative growth	[55,71]
CsPAO6	Cs7g23760.1	n.d.	n.d.	BC (?)	stem, root, cotyledon	root growth, vegetative growth	[55,71]

n. d., not determined; Br, breaker stage fruit; Br+2, two days post breaker stage fruit; Br+7, seven days post breaker stage fruit; BC, back conversion; TC, terminal catabolism.

Above all, two different kinds of PAOs exist in rice; one is BC-type (OsPAO1, OsPAO3~5), the other is TC-type (OsPAO7, and OsPAO6 possibly also has this activity), as shown in Table 2 [46,49,60]. To fully understand the biological functions of *OsPAOs* in various developmental and physiological processes, molecular and genetic approaches like CRISPR/Cas9-mediated loss-of-function mutants and ubiquitin promoter enhanced overexpression transgenic plants should be generated.

4.2. Arabidopsis PAOs

The Arabidopsis genome contains five *PAOs*, named as *AtPAO1* to *AtPAO5*. The recombinant protein of the former four *AtPAOs*, AtPAO1~4, have been homogenously purified and characterized [12,39,61,62,64,65]; besides, AtPAO5 also has been purified and biochemically characterized [46,51]. In detail, AtPAO1, subcellularly localized in cytoplasm, catalyzes a BC-type reaction, and prefers to utilize Spm, T-Spm, and NorSpm as substrates [39]; AtPAO2~4, localized to peroxisomes, all display a BC-type reaction with different substrate specificity [12,61,62,64]. AtPAO2~3 oxidize Spm to Put in a full BC-type reaction via Spd, whereas the other peroxisomal AtPAO4 mainly catalyzes the partial BC-type because only very few Put can be detected when Spm was used as the substrate [61].

Five Arabidopsis *PAOs* showed different expression patterns. *AtPAO1* is specifically expressed in the root transition region (between the meristematic and elongation zones of the root) and anther tapetum [65], and Takahashi et al. also found that *AtPAO1* is specifically expressed in anthers [62]. *AtPAO1* was reported to be involved in environment stress tolerance [39,65], and the expression patterns imply *AtPAO1* may also play roles in root development and fertility, as shown in Table 2 [62]. *AtPAO2* is mainly expressed in the root and shoot meristematic area, the vein of rosette leaves, as well as the anthers, suggesting that *AtPAO2* might function in the development of roots, shoots, leaves, and flowers, as shown in Table 2 [62]. *AtPAO3* and *AtPAO4* display similar expression patterns, which are expressed in all tissues and whole growth stages, especially in roots, leaves, and flowers, suggesting that these two members may mediate various significant growth processes, as shown in Table 2 [62]. *pao4-1* and *pao4-2*, two independent lines of *AtPAO4* loss-of-function mutants, have 10-fold higher Spm levels compared to wild type, and delay dark-triggered senescence [66]. The last Arabidopsis PAO, *AtPAO5*, is expressed in all developmental stages, with strongest expression in roots, stems, leaves, and floral organs, as shown in Table 2 [51,62].

AtPAO5 is a relatively completely explained Arabidopsis PAO, and its gene product AtPAO5 has been successfully characterized and its biological function also has been explored [51,67,68]. AtPAO5 can catalyze both Spm and T-Spm to Spd, but not to Put [51]. Our former colleagues Kim et al. reported that *AtPAO5* regulates stem elongation and the rosette leaves' development, as shown in Table 2 [51,62]. Two *AtPAO5* T-DNA insertion mutants, *pao5-1* and *pao5-2*, both of which show about 2-fold higher levels of T-Spm, still maintain normal levels of Put, Spd, and Spm compared to the wild type controls [51]. The *pao5-1* and *pao5-2* mutants exhibit more rosette leaves, and shorter and fewer inflorescence stems at the two-month-old stage. Further genetic and morphology analysis suggested that *AtPAO5* plays roles in Arabidopsis growth and development through oxidizing T-Spm [46,51]. Ahou et al. found that AtPAO5 functions as an SMO/dehydrogenase [69]. *atpao5-2* and *atpao5-3*, two independent loss-of-function mutants of *AtPAO5*, show higher T-Spm contents, mediate metabolic and transcriptional reprogramming, and enhance salt-related stress tolerance [67]. *AtPAO5* also plays roles in the control of proper xylem differentiation through interplaying between auxin and cytokinins [68]. Above all, the *AtPAO5* mutant with higher T-Spm levels shows the similar phenotypes as *acl5* (*tkv*) and *bud2* mutants, which only contain very low or even zero T-Spm content [7,9,15,62,72,73]. These results explained that maintaining suitable T-Spm content is very important in plants.

Taken together, all five Arabidopsis PAOs catalyze BC-type reactions and mediate (or potentially mediate) the entire developmental processes in plants, as shown in Figure 2 and Table 2 [46], and their (especially the *AtPAO1~4*) biological functions need to be further unveiled in the future.

4.3. Tomato PAOs

Transgenic tomato plants overexpressing maize PAO (MPAO) exhibit tissue damage with lower chlorophyll content, lower photochemical efficiency of photosystem II (PSII), and DNA fragmentation compared to wild type, suggesting that the increased PAO activity cannot cope with the reactive oxygen species (ROS) generated by environmental factors [13]. In *S. lycopersicum cv.* Chiou, the expression of PAO peaked at ImG1 (fruits 0.5 cm in diameter) and ImG2 (fruits 1 cm in diameter) stages, suggesting PAO participates in developmental processes of the fruits, including the cell wall maturation [74]. Gémes et al. reported that sense-*ZmPAO* (*S-ZmPAO*) transgenic tomato plants have slightly larger leaf sizes and higher antioxidant enzyme activities; in contrast, the antisense-*ZmPAO* (*AS-ZmPAO*) transgenic tomato plants contain lower chlorophyll content index, smaller leaves, and less biomass, as well as an increment in Ca^{2+} when responding to salt stress [29]. The phenotypes of *S-ZmPAO* and *AS-ZmPAO* transgenic plants suggested that apoplastic PAO play important roles in plant growth and stress responses [29]. Most recently, we found that the model dicotyledons of the tomato plant (*Solanum lycopersicum*) has seven *PAO* genes in its genome, which were orderly named as *SlPAO1* to *SlPAO7* [63]. *SlPAO2~5*, sharing high identity (over 64%) of amino acid and showing quite similar genome organization and predicted tertiary structures, have similar tissue expression patterns [63]. Besides, *SlPAO2~4* are ubiquitously and highly expressed in the whole growth processes and all tissues, predominantly in anther, Br (breaker stage fruit), and Br+2 (two days post breaker stage fruit) [63], suggesting that *SlPAO2~4* may play dominant roles in all stages of growth especially in floral development and fruit maturity in tomato, as shown in Table 2 [63]. *SlPAO1* is expressed relatively lower than *SlPAO2~4* in all of the vegetative tissues and anthers [63]. What is more, *SlPAO6~7*, sharing quite similar identity of amino acid and very similar intron-exon organization and protein 3-D structures, are lowly expressed in vegetative and reproductive tissues, but had relatively higher expression in roots, stems, buds, and anthers than in the fruit [63], suggesting that these two tomato *PAOs* may mainly function in vegetative and anthesis tissues but not in fruit, as shown in Table 2. *SlPAOs* respond to abiotic stresses (heat, wound, cold, drought, and salt), oxygen species (H_2O_2 and methylviologen), phytohormones (IAA, 6-BA, GA, ABA, Eth, SA, and JA), as well as PAs (Put, Spd, Spm, and T-Spm), implying that tomato *PAOs* possibly have various functions in stress tolerances, as shown in Table 2 [63]. Taken together, *SlPAOs* possibly play vital roles in different tissues and developmental stages, especially in floral development and fruit repining. To better explain the mechanism of polyamine catabolism and biological roles of *SlPAOs*, more biochemical and genetic experiments are required.

4.4. PAOs in Other Plants

Recently, besides these three model plants (rice, Arabidopsis, and tomato), some other plant species have also been studied on PAO catabolism, and PAO biological functions. Plant PAOs play important roles in various stress tolerance and the programmed cell death (PCD) events through mediating H_2O_2 signaling which is generated by stress-induced PAO activity leading to Spd, Spm, and T-Spm oxidation [13,33,75–98]. Hatmi et al. reported that the grapevine PAO and CuAO activities were upregulated by osmotic stress and *Botrytis cinerea* infection, suggesting that PA back-conversion and/or terminal catabolism were involved in PA homeostasis under stress conditions [97]. In addition, the PAO activity increment and proline accumulation were involved in cold tolerance in *Medicago falcate* [75,76], suggesting that PAOs and proline interplay in the process of various stress responses [75,76,99–101]. What is more, in *salinity tolerance 1 (st1)*, a wheat salinity-tolerant line, the expression of *PAO* genes showed high expression levels, suggesting that *PAO* genes may have important functions in salinity tolerance [102].

Previously, Sagor and his colleagues reported that SelPAO5 from *Selaginella lepidophylla* back-converts Spm and T-Spm to Spd and Nor-Spd, respectively [53]. It is different from AtPAO5 and OsPAO1 which prefer to use the same substrates as SelPAO5, but both of these two enzymes convert the substrates to Spd, though three of them are from the same clade in the phylogenic relationship tree, suggesting that SelPAO5 oxidizes T-Spm at different carbon positions [53]. Most recently, they further found that the SelPAO5 can complement the dwarf phenotype of *Atpao5*, with the reduction of T-Spm content to almost normal levels of wild type, which strengthens the claim that T-Spm homeostasis is required for plant development and growth [103]. Besides, Wang and Liu firstly identified PAOs from sweet orange (*Citrus sinensis*); their results indicated that six PAO genes (*CsPAO1*–*CsPAO6*) exist in sweet orange, and they also found that CsPAO3 may have potential roles in PA back conversion in plants, while CsPAO4 catalyzes Spd and Spm as substrates for terminal catabolism [55,71]. The transgenic plants overexpressing *CsPAO4* showed growth inhibition under salt stress caused by the elevation of H_2O_2 which leads to oxidative damages [55]. What is more, Brikis et al. found that the expression of *MdPAO2* was obviously upregulated in apple fruit by elevating the CO_2 concentrations under low-temperature/low-O_2 storage for up to sixteen weeks, suggesting that *MdPAO2* is involved in respiratory activities in apple fruit storage under multiple abiotic stresses [104]. Furthermore, Takahashi et al. characterized the molecular and biochemical features of five PAOs (BdPAO1 to BdPAO5) from *Brachypodium distachyon*, and they found that BdPAO2 and BdPAO4 possibly are localized to peroxisomes [70]. Additionally, they also found that BdPAO2 catalyzes a full-back conversion pathway, and the favorite substrates of BdPAO2 and BdPAO3 are Spd and Spm, respectively [70].

Plant PAOs play significant roles in metal toxicity tolerance. Aluminum (Al), copper (Cu), and cadmium (Cd), etc. are phytotoxic to plants at high concentrations [33,81,97,105,106]. In wheat, the cell wall-bound PAO (CW-PAO) oxidized Spd and generated H_2O_2 under Al toxicity; in contrast, the CW-PAO activity was markedly inhibited by Put application, and subsequently reduced H_2O_2 accumulation in roots under Al stress, suggesting that Put plays an important protective role against Al-induced oxidative stress via inhibiting the PAO activity with lower H_2O_2 production [33]. Similarly, the PAO activity was enhanced by higher Cu or Cd concentrations leading to accelerating the PA back-conversion or terminal catabolism, which may be related to functionality of defense mechanisms [105,106]. To entirely understand the functional mechanism of PAOs on metal toxicity tolerance, more attractive and systematic studies are required.

Plant PAOs have important roles in plant growth and development. Around fifteen years ago, the functions of the maize PAO were investigated by the Rea group and the Cona group separately, and they found that the maize PAO plays roles in cell-wall maturation and root differentiation by producing H_2O_2 [107,108]. Gomez-Jimenez et al. reported that PAO and DAO have significant functions in olive fruit abscission zone (AZ) development through providing apoplastic H_2O_2 for cell-wall strengthening and lignosuberization events, and the peroxidase substrate is provided in these cells throughout AZ development [109]. Moreover, Rodríguez et al. reported that the increased PAO activity produces more H_2O_2 to generate $\cdot O^{2-}$ through enhanced substrate availability and subsequently maintain maize leaf elongation under saline stress [16]. What is more, the tomato PAO is involved in vascular development via mediating H_2O_2 which is required by vascular differentiation and the process of polymerization of lignin precursors into lignin [110]. *Atpao3*, a loss-of-function mutant of AtPAO3 which oxidizes Spd in peroxisomes [12], shows reduced pollen tube and seed setting caused by significantly disrupted Spd-induced Ca^{2+} currents [111]. Furthermore, Agudelo-Romero et al. found that the activities of PAO and DAO are significantly increased during grape ripening, implying an important role of polyamines' catabolism in fruit ripening [112].

4.5. Peroxisomal PAOs in Plants

In Arabidopsis, *AtPAO2~4* were speculated to be localized to peroxisomes [12,39,62,64,65]; additionally, in rice, we also found that *OsPAO3~5* are situated in peroxisomes [46,47]. Besides, recently some other groups reported that *BdPAO2* and *BdPAO4* from *Brachypodium distachyon* [70], *BrPAO2~4* from *Brassica rapa* [81], *CsPAO2~3* from *Citrus sinensis* [71], and *SlPAO2~4* from tomato [63] were predicted to be peroxisomal PAOs. All of these genes' products classifying into clade IV, as shown in Figure 2, contain peroxisomal-targeting signals in their C-terminal, resulting in localization to peroxisome, as shown in Figure 3 [12,39,46,47,62–65,70,81]. In the apple genome, six putative apple PAO genes were identified [104]. The *MdPAO2~4* were predicted to localize in peroxisomes, whereas *MdPAO1* and *MdPAO5~6* were predicted to be cytosolic proteins [104]. In addition, four CuAO-like genes from *Arabidopsis* have two different localizations; the AtCuAO2 and AtCuAO3 are localized to peroxisomes, while the AtAO1 and AtCuAO1 are localized to apoplast [113].

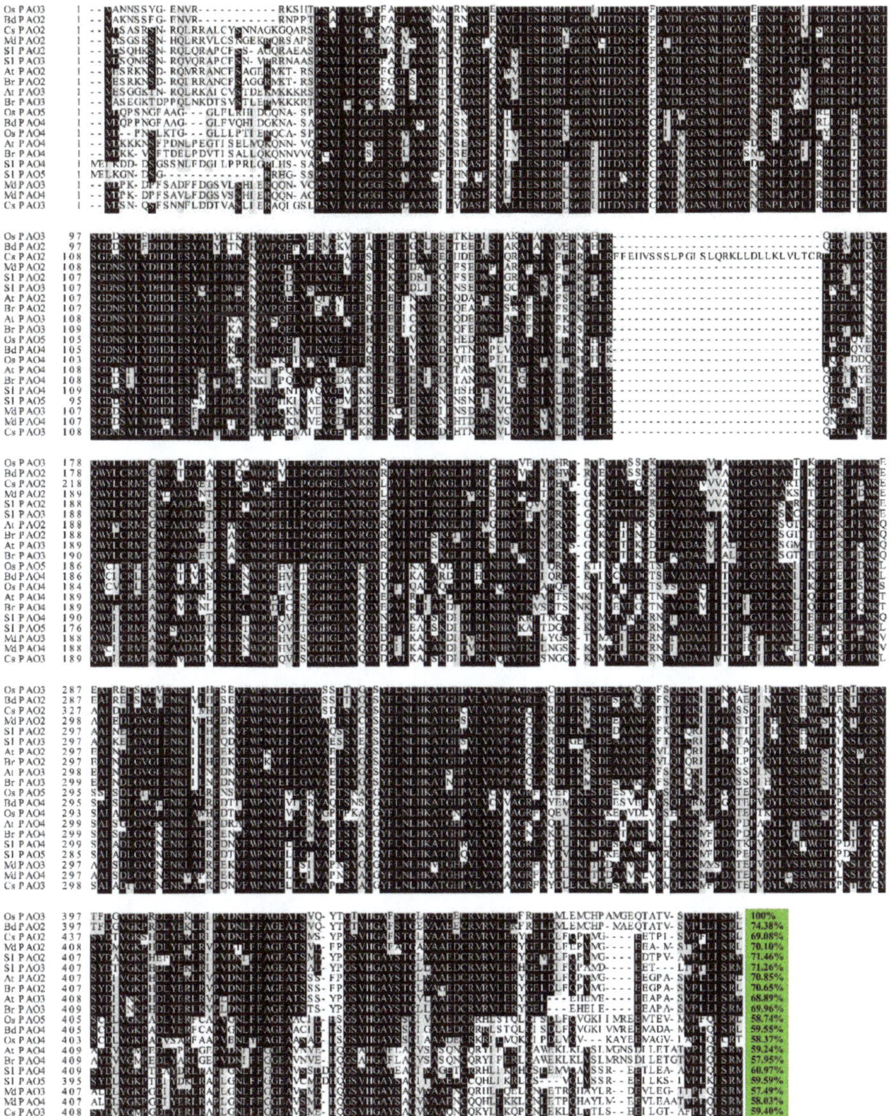

Figure 3. Alignment of amino acid sequences of twenty reported peroxisomal PAOs from *Oryza sativa*, *Arabidopsis thaliana*, *Solanum lycopersicum*, *Brachypodium distachyon*, *Brassica rapa*, *Citrus sinensis*, and *Malus domestica*. The alignment was performed by the Clustal X 1.83 software and exhibited by the Boxshade program (http://www.ch.embnet.org/software/BOX_form.html.). Black and gray indicate the complete and partial homology of the amino acid sequences, respectively. The percentages at the end of the alignment showed the identity between OsPAO3 and other PAOs.

These peroxisomal PAOs shared high identity (over 57% compared to OsPAO3 which was set as 100%), as shown in Figure 3, and displayed quite similar predicted protein tertiary structures, as shown in Figure 4A–Q, even though these PAOs are from six different species. Interestingly, the predicted protein tertiary structures of these twenty peroxisomal PAOs almost fully merged with each other, as shown in Figure 4U, except OsPAO4 and CsPAO2 that cannot merge with other PAOs; whereas, to our surprise, the OsPAO4 and CsPAO2 were largely merged, as shown in Figure 4V. Besides, the protein sequence of CsPAO2 contains an additional twenty-nine amino acid sequence in the conserved region compared to other peroxisomal PAOs, as shown in Figure 3, that may be because of the mRNA alternative splicing, though the possible function of this additional sequence remains totally unknown. The results of phylogenetic relationship analysis, as shown in Figure 4W, also indicated that these peroxisomal PAOs are highly conserved and extremely close during evolution in the plant kingdom.

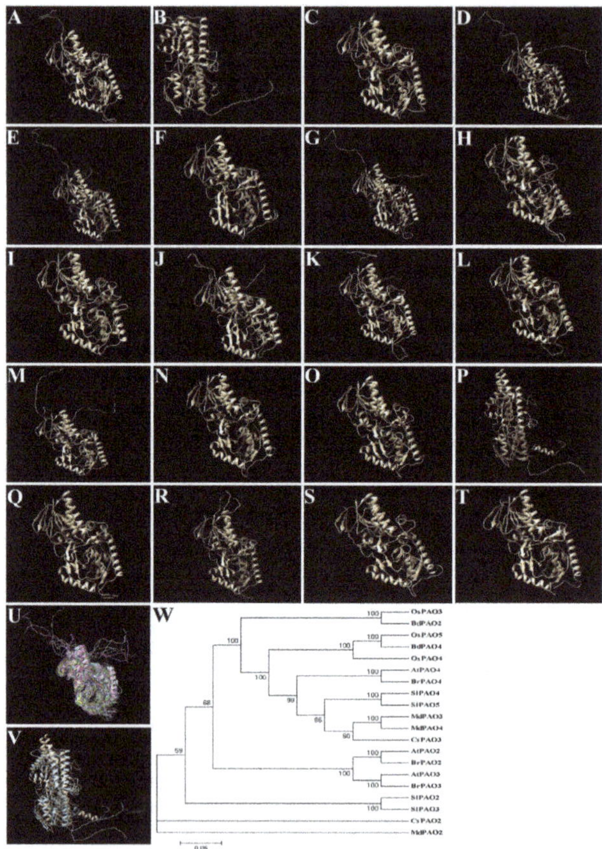

Figure 4. Predicted tertiary structures of the reported peroxisomal PAOs. Twenty have been reported; peroxisomal plant PAOs were analyzed. (**A–J**), The protein 3-D structures of OsPAO3 (**A**); OsPAO4 (**B**); OsPAO5 (**C**); AtPAO2 (**D**); AtPAO3 (**E**); AtPAO5 (**F**); SlPAO2 (**G**); SlPAO3 (**H**); SlPAO4 (**I**); SlPAO5 (**J**); BdPAO2 (**K**); BdPAO4 (**L**); BrPAO2 (**M**); BrPAO3 (**N**); BrPAO4 (**O**); CsPAO2 (**P**); CsPAO3 (**Q**); MdPAO2 (**R**); MdPAO3 (**S**); and MdPAO4 (**T**) were obtained using the Protein Structure Prediction Server program (http://ps2v3.life.nctu.edu.tw/) and Chimera 1.13 software. (**U**) Merged image of all PAOs, except OsPAO4 and CsPAO2, was performed by Chimera 1.13 software. (**V**) Merged image of OsPAO4 and CsPAO2 was similarly performed. The light blue and light yellow colors indicate the protein structures of OsPAO4 and CsPAO2, respectively. (**W**) Evolution relationship among the peroxisomal PAOs.

It is suggested that the peroxisomal PAOs possibly play significant roles in plant growth processes, especially in floral development, as shown in Table 2. To explore the physiological and biological significance of peroxisomal PAOs, genetic and morphological approaches are required via generating functional knock-down (or knock-out) mutants. Besides, apoplastic PAOs were found in monocotyledonous plants such as maize PAO (ZmPAO), barley PAO (HvPAOs), and rice PAO (OsPAOs), which were involved in TC-type pathways to catalyze PA terminal oxidation [36,42,44–46,49,60,107]. In dicots, apoplastic PAOs may be present in limited species [55]. What is more, the cytoplasmic PAOs were characterized in Arabidopsis (AtPAO1 and AtPAO5) [39,51,61,62,65] and rice (OsPAO1) [46,48], which catalyzed PA back conversion reactions. However, the roles of the three types of PAOs in plant growth and development, and stress tolerance through PA homeostasis and/or H_2O_2 generation, remain fragmentary. Thereby, the significance of the functional difference between peroxisomal or cytoplasmic PAOs and apoplastic PAOs remains to be clarified and should be addressed in future work.

5. Conclusions and Future Perspective of *PAOs* Research in Plants

In the past years, some PAO genes were cloned and functionally identified from different plant species. Some research groups focus on the PA catabolism pathway, meanwhile, more and more researchers pay intense attention to the biological roles of PAOs. As it is known, when plants grow under normal conditions, the intracellular PAs maintain homeostasis, and the normal level of H_2O_2 is generated by PAOs. Subsequently the H_2O_2 signal participates in the developmental processes such as root growth, xylem differentiation, pollen tube growth, fruit development, etc., as shown in Figure 5 [65,69,99,111,112]. However, the PAs homeostasis might encounter challenges under stress conditions. The enhanced accumulation of stress-induced PAs requires higher PAO activity to rebalance the PAs homeostasis. If just under mild stress, the plants can overcome the unpleasant period via the antioxidant reaction with the aid of proline and other catabolites that were also induced by stress [76,99–101]. If under severe stress and longtime stress conditions, the PAO activity markedly increases to reduce the stress-induced intracellular PAs level with high H_2O_2 accumulation, leading to a ROS burst which may result in death, as shown in Figure 5. The antioxidant activity cannot offset the strong ROS burst, though the levels of proline and other catabolites are also upregulated by stress, as shown in Figure 5 [93,94].

Recently, the enzyme features of all the Arabidopsis *PAOs* and most rice *PAOs* have been identified, but their biological roles remain largely unclear. Meanwhile, the tomato *PAOs* have been cloned, but its catabolic activities and biological functions are still unknown. What is more, the exact roles of the highly conserved peroxisomal *PAOs* in plants are still fuzzy. Furthermore, why does rice have two different types of *PAO* catabolic pathways (the BC-type and TC-type pathway)? In addition, the exact mechanism of PA metabolism and the PA-cycle—PA exodus—as well as the possible ratio between the back-conversion and terminal catabolism in plants needs to be uncovered. Finally, what is the possible relationship between PAO and proline when plants fight against environmental stresses? To fully understand the roles of *PAOs* in plant development and stress interactions, intensive studies are required via generating loss-of-function mutants and overexpression transgenic plants which will greatly help further explore the biochemical and physiological roles of these *PAOs*.

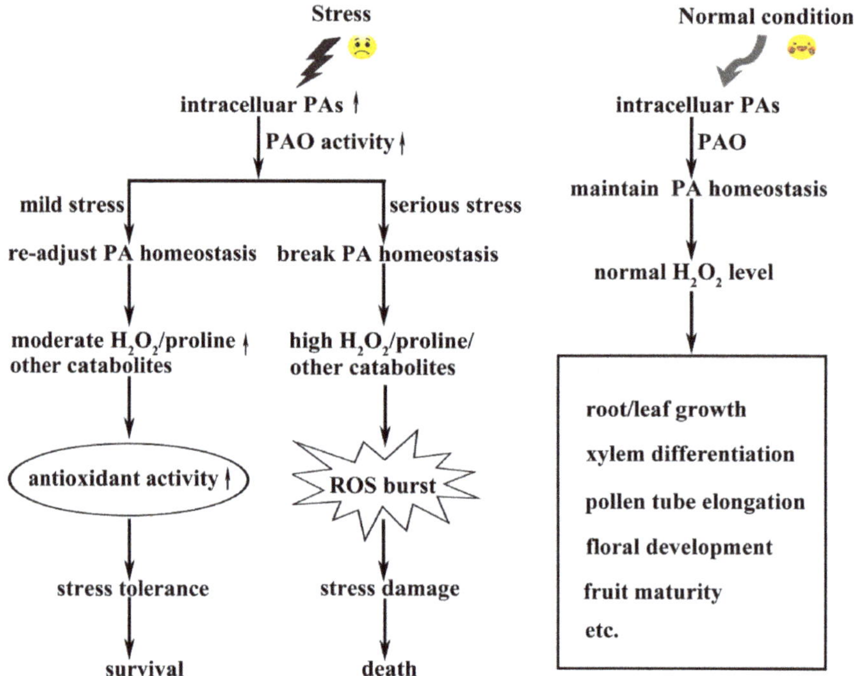

Figure 5. Diagrammatic representation of the roles of PAO involved in developmental growth and environmental stress response in plants. The thick upright arrows indicate increase in the activity or concentrations. The cartoon pictures of smiling and bitter faces indicate the plant growth under normal or stress conditions, respectively. ROS: reactive oxygen species.

Author Contributions: Z.Y. analyzed the data and made the figures; D.J. revised the manuscript; T.L. designed the study, analyzed the data, and wrote the manuscript. All authors read and approved the final manuscript.

Funding: This research was funded by National Natural Science Foundation of China (31600217).

Conflicts of Interest: The authors have no conflicts of interest to declare.

References

1. Cohen, S.S. *A Guide to the Polyamines*; Oxford University Press: New York, NY, USA, 1998.
2. Tabor, C.W.; Tabor, H. Polyamines. *Annu. Rev. Plant. Biol.* **1984**, *53*, 749–790. [CrossRef] [PubMed]
3. Knott, J.M.; Römer, P.; Sumper, M. Putative spermine synthases from *Thalassiosira pseudonana* and *Arabidopsis thaliana* synthesize thermospermine rather than spermine. *FEBS Lett.* **2007**, *581*, 3081–3086. [CrossRef] [PubMed]
4. Kusano, T.; Berberich, T.; Tateda, C.; Takahashi, Y. Polyamines, essential factors for growth and survival. *Planta* **2008**, *228*, 367–381. [CrossRef] [PubMed]
5. Handa, A.K.; Mattoo, A. Differential and functional interactions emphasize the multiple roles of polyamines in plants. *Plant Physiol. Biochem.* **2010**, *48*, 540–546. [CrossRef]
6. Mattoo, A.K.; Minocha, S.C.; Minocha, R.; Handa, A.K. Polyamines and cellular metabolism in plants, transgenic approaches reveal different responses to diamine putrescine versus higher polyamines spermidine and spermine. *Amino Acids* **2010**, *38*, 405–413. [CrossRef] [PubMed]
7. Takano, A.; Kakehi, J.; Takahashi, T. Thermospermine is not a minor polyamine in the plant kingdom. *Plant Cell Physiol.* **2012**, *53*, 606–616. [CrossRef] [PubMed]
8. Hanzawa, Y.; Takahashi, T.; Komeda, Y. ACL5, an Arabidopsis gene required for internodal elongation after flowering. *Plant J.* **1997**, *12*, 863–874. [CrossRef] [PubMed]

9. Hanzawa, Y.; Takahashi, T.; Michael, A.J.; Burtin, D.; Long, D.; Pineiro, M.; Coupland, G.; Komeda, Y. *ACAULIS5*, an Arabidopsis gene required for stem elongation, encodes a spermine synthase. *EMBO J.* **2000**, *19*, 4248–4256. [CrossRef] [PubMed]

10. Bouchereau, A.; Aziz, A.; Larher, F.; Martin, T.J. Polyamines and environmental challenges, recent development. *Plant Sci.* **1999**, *140*, 103–125. [CrossRef]

11. Minocha, R.; Majumdar, R.; Minocha, S.C. Polyamines and abiotic stress in plants: A complex relationship. *Front. Plant Sci.* **2014**, *5*, 175. [CrossRef]

12. Moschou, P.N.; Sanmartin, M.; Andriopoulou, A.H.; Rojo, E.; Sanchez-Serrano, J.J.; Roubelakis-Angelakis, K.A. Bridging the gap between plant and mammalian polyamine catabolism, a novel peroxisomal polyamine oxidase responsible for a full back-conversion pathway in Arabidopsis. *Plant Physiol.* **2008**, *147*, 1845–1857. [CrossRef] [PubMed]

13. Moschou, P.N.; Paschalidis, K.A.; Delis, I.D.; Andriopoulou, A.H.; Lagiotis, G.D.; Yakoumakis, D.I.; Roubelakis-Angelakis, K.A. Spermidine exodus and oxidation in the apoplast induced by abiotic stress is responsible for H_2O_2 signatures that direct tolerance responses in tobacco. *Plant Cell.* **2008**, *20*, 1708–1724. [CrossRef] [PubMed]

14. Moschou, P.N.; Roubelakis-Angelakis, K.A. Polyamines and programmed cell death. *J. Exp. Bot.* **2014**, *65*, 1285–1296. [CrossRef] [PubMed]

15. Kakehi, J.I.; Kuwashiro, Y.; Niitsu, M.; Takahashi, T. Thermospermine is required for stem elongation in *Arabidopsis thaliana*. *Plant Cell Physiol.* **2008**, *49*, 1342–1349. [CrossRef] [PubMed]

16. Rodriguez, A.A.; Maiale, S.J.; Menendez, A.B.; Ruiz, O.A. Polyamine oxidase activity contributes to sustain maize leaf elongation under saline stress. *J. Exp. Bot.* **2009**, *60*, 4249–4262. [CrossRef] [PubMed]

17. Naka, Y.; Watanabe, K.; Sagor, G.H.M.; Niitsu, M.; Pillai, M.A.; Kusano, T.; Takahashi, Y. Quantitative analysis of plant polyamines including thermospermine during growth and salinity stress. *Plant Physiol. Biochem.* **2010**, *48*, 527–533. [CrossRef] [PubMed]

18. Groppa, M.D.; Benavides, M.P. Polyamines and abiotic stress, recent advances. *Amino Acids* **2008**, *34*, 35–45. [CrossRef] [PubMed]

19. Alcazar, R.; Altabella, T.; Marco, F.; Bortolotti, C.; Reymond, M.; Koncz, C.; Carrasco, P.; Tiburcio, A.F. Polyamines, molecules with regulatory functions in plant abiotic stress tolerance. *Planta* **2010**, *231*, 1237–1249. [CrossRef] [PubMed]

20. Marina, M.; Sirera, F.V.; Rambla, J.L.; Gonzalez, M.E.; Blazquez, M.A.; Carbonell, J.; Pieckenstain, F.L.; Ruiz, O.A. Thermospermine catabolism increases *Arabidopsis thaliana* resistance to *Pseudomonas viridiflava*. *J. Exp. Bot.* **2013**, *64*, 1393–1402. [CrossRef] [PubMed]

21. Hatmi, S.; Gruau, C.; Trotel-Aziz, P.; Villaume, S.; Rabenoelina, F.; Baillieul, F.; Eullaffroy, P.; Clément, C.; Ferchichi, A.; Aziz, A. Drought stress tolerance in grapevine involves activation of polyamine oxidation contributing to improved immune response and low susceptibility to *Botrytis cinerea*. *J. Exp. Bot.* **2015**, *66*, 775–787. [CrossRef] [PubMed]

22. Cai, G.; Sobieszczuk-Nowicka, E.; Aloisi, I.; Fattorini, L.; Serafini-Fracassini, D.; Del Duca, S. Polyamines are common players in different facets of plant programmed cell death. *Amino Acids* **2015**, *47*, 27–44. [CrossRef] [PubMed]

23. Liu, J.H.; Wang, W.; Wu, H.; Gong, X.Q.; Moriguchi, T. Polyamines function in stress tolerance, from synthesis to regulation. *Front. Plant Sci.* **2015**, *6*, 827. [CrossRef] [PubMed]

24. Huang, X.-S.; Zhang, Q.; Zhu, D.; Fu, X.; Wang, M.; Zhang, Q.; Moriguchi, T.; Liu, J.-H. *ICE1* of *Poncirus trifoliata* functions in cold tolerance by modulating polyamine levels by interacting with arginine decarboxylase. *J. Exp. Bot.* **2015**, *68*, 3259–3274. [CrossRef] [PubMed]

25. Ebeed, H.T.; Hassan, N.M.; Aljarani, A.M. Exogenous applications of Polyamines modulate drought responses in wheat through osmolytes accumulation, increasing free polyamine levels and regulation of polyamine biosynthetic genes. *Plant Physiol. Biochem.* **2017**, *118*, 438–448. [CrossRef] [PubMed]

26. Yang, W.B.; Li, Y.; Yin, Y.P.; Qin, Z.L.; Zheng, M.J.; Chen, J.; Luo, Y.L.; Pang, D.W.; Jiang, W.W.; Li, Y.; et al. Involvement of ethylene and polyamines biosynthesis and abdominal phloem tissues characters of wheat caryopsis during grain filling under stress conditions. *Sci. Rep.* **2017**, *7*, 46020. [CrossRef]

27. Park, J.Y.; Kang, B.R.; Ryu, C.M.; Anderson, A.J.; Kim, Y.C. Polyamine is a critical determinant of Pseudomonas chlororaphis O6 for GacS-dependent bacterial cell growth and biocontrol capacity. *Mol. Plant Pathol.* **2018**, *19*, 1257–1266. [CrossRef] [PubMed]

28. Wuddineh, W.; Minocha, R.; Minocha, S.C. *Polyamines in the Context of Metabolic Networks*; Alcázar, R., Tiburcio, A., Eds.; Humana Press: New York, NY, USA, 2018; pp. 1–23.

29. Gémes, K.; Mellidou, I.; Karamanoli, K.; Beris, D.; Park, K.Y.; Matsi, T.; Haralampidis, K.; Constantinidou, H.I.; Roubelakis-Angelakis, K.A. Deregulation of apoplastic polyamine oxidase affects development and salt response of tobacco plants. *J. Plant Physiol.* **2017**, *211*, 1–12. [CrossRef] [PubMed]

30. Mellidou, I.; Karamanoli, K.; Beris, D.; Haralampidis, K.; Constantinidou, H.A.; Roubelakis-Angelakis, K.A. Underexpression of apoplastic polyamine oxidase improves thermotolerance in *Nicotiana tabacum*. *J. Plant Physiol.* **2017**, *218*, 171–174. [CrossRef] [PubMed]

31. Chen, B.-X.; Li, W.-Y.; Gao, Y.-T.; Chen, Z.-J.; Zhang, W.-N.; Liu, Q.-J.; Chen, Z. Involvement of polyamine oxidase-produced hydrogen peroxide during coleorhiza-limited germination of rice seeds. *Front. Plant Sci.* **2016**, *7*, 1219. [CrossRef] [PubMed]

32. Sobieszczuk-Nowicka, E. Polyamine catabolism adds fuel to leaf senescence. *Amino Acids* **2017**, *49*, 49–56. [CrossRef] [PubMed]

33. Yu, Y.; Zhou, W.W.; Zhou, K.J.; Liu, W.J.; Liang, X.; Chen, Y.; Sun, D.; Lin, X. Polyamines modulate aluminum-induced oxidative stress differently by inducing or reducing H_2O_2 production in wheat. *Chemosphere* **2018**, *212*, 645–653. [CrossRef] [PubMed]

34. Bagni, N.; Tassoni, A. Biosynthesis, oxidation and conjugation of aliphatic polyamines in higher plants. *Amino Acids* **2001**, *20*, 301–317. [CrossRef] [PubMed]

35. Wallace, H.M.; Fraser, A.V.; Hughes, A. A perspective of polyamine metabolism. *Biochem. J.* **2003**, *376*, 1–14. [CrossRef] [PubMed]

36. Federico, R.; Cona, A.; Angelini, R.; Schininà, M.E.; Giartosio, A. Characterization of maize polyamine oxidase. *Phytochemistry* **1990**, *29*, 2411–2414. [CrossRef]

37. Federico, R.; Ercolini, L.; Laurenzi, M.; Angelini, R. Oxidation of cetylpolyamines by maize polyamine oxidase. *Phytochemistry* **1996**, *43*, 339–341. [CrossRef]

38. Tavladoraki, P.; Shinina, M.E.; Cecconi, F.; Di Agostino, S.; Manera, F.; Rea, G.; Mariottini, P.; Federico, R.; Angelini, R. Maize polyamine oxidase, primary structure from protein and cDNA sequencing. *FEBS Lett.* **1998**, *426*, 62–66. [CrossRef]

39. Tavladoraki, P.; Rossi, M.N.; Saccuti, G.; Perez-Amador, M.A.; Polticelli, F.; Angelini, R.; Federico, R. Heterologous expression and biochemical characterization of a polyamine oxidase from Arabidopsis involved in polyamine back conversion. *Plant Physiol.* **2006**, *141*, 1519–1532. [CrossRef]

40. Tavladoraki, P.; Cona, A.; Angelini, R. Copper-containing Amine oxidases and FAD-dependent polyamine oxidases are key players in plant tissue differentiation and organ development. *Front. Plant Sci.* **2016**, *7*, 824. [CrossRef]

41. Radova, A.; Sebela, M.; Galuszka, P.; Frebort, I.; Jacobsen, S.; Faulhammer, H.G.; Pec, P. Barley polyamine oxidase, characterisation and analysis of the cofactor and the N-terminal amino acid sequence. *Phytochem. Anal.* **2001**, *12*, 166–173. [CrossRef]

42. Cervelli, M.; Cona, A.; Angelini, R.; Polticelli, F.; Federico, R.; Mariottini, P. A barley polyamine oxidase isoform with distinct structural features and subcellular localization. *Eur. J. Biochem.* **2001**, *268*, 3816–3830. [CrossRef]

43. Cervelli, M.; Polticelli, F.; Federico, R.; Mariottini, P. Heterologous expression and characterization of mouse spermine oxidase. *J. Biol. Chem.* **2003**, *278*, 5271–5276. [CrossRef] [PubMed]

44. Cervelli, M.; Di Caro, O.; Di Penta, A.; Angelini, R.; Federico, R.; Vitale, A.; Mariottini, P. A novel C-terminal sequence from barley polyamine oxidase is a vacuolar sorting signal. *Plant J.* **2004**, *40*, 410–418. [CrossRef] [PubMed]

45. Cervelli, M.; Bianchi, M.; Cona, A.; Crosatti, C.; Stanca, M.; Angelini, R.; Federico, R.; Mariottini, P. Barley polyamine oxidase isoforms 1 and 2, a peculiar case of gene duplication. *FEBS J.* **2006**, *273*, 3990–4002. [CrossRef] [PubMed]

46. Kusano, T.; Kim, D.; Liu, T.; Berberich, T. *Polyamine Catabolism in Plants*; Springer: Tokyo, Japan, 2015.

47. Ono, Y.; Kim, D.W.; Watanabe, K.; Sasaki, A.; Niitsu, M.; Berberich, T.; Kusano, T.; Takahashi, Y. Constitutively and highly expressed *Oryza sativa* polyamine oxidases localize in peroxisomes and catalyze polyamine back conversion. *Amino Acids* **2012**, *42*, 867–876. [CrossRef] [PubMed]

48. Liu, T.; Kim, D.W.; Niitsu, M.; Berberich, T.; Kusano, T. *Oryza sativa* polyamine oxidase 1 back-converts tetraamines, spermine and thermospermine, to spermidine. *Plant Cell Rep.* **2014**, *33*, 143–151. [CrossRef]

49. Liu, T.B.; Kim, D.W.; Niitsu, M.; Maeda, S.; Watanabe, M.; Kamio, Y.; Berberich, T.; Kusano, T. Polyamine oxidase 7 is a terminal catabolism-type enzyme in *Oryza sativa* and is specifically expressed in anthers. *Plant Cell Physiol.* **2014**, *55*, 1110–1122. [CrossRef]

50. Liu, T.B.; Dobashi, H.; Kim, D.W.; Sagor, G.H.M.; Niitsu, M.; Berberich, T.; Kusano, T. Differential sensitivity to exogenous cadaverine of Arabidopsis mutants with defect in polyamine metabolic genes may be explained by spermine content in their plants. *Physiol. Mol. Biol. Plants* **2014**, *20*, 151–159. [CrossRef]

51. Kim, D.W.; Watanabe, K.; Murayama, C.; Izawa, S.; Niitsu, M.; Michael, A.J.; Berberich, T.; Kusano, T. Polyamine oxidase 5 regulates Arabidopsis growth through thermospermine oxidase activity. *Plant Physiol.* **2014**, *165*, 1575–1590. [CrossRef]

52. Mo, H.J.; Wang, X.F.; Zhang, Y.; Zhang, G.Y.; Zhang, J.F.; Ma, Z.Y. Cotton polyamine oxidase is required for spermine and camalexin signalling in the defence response to *Verticillium dahlia*. *Plant J.* **2015**, *83*, 962–975. [CrossRef]

53. Sagor, G.H.M.; Inoue, M.; Kim, D.W.; Kojima, S.; Niitsu, M.; Berberich, T.; Kusano, T. The polyamine oxidase from lycophyte *Selaginella lepidophylla* (SelPAO5), unlike that of angiosperms, back-converts thermospermine to norspermidine. *FEBS Lett.* **2015**, *589*, 3071–3078. [CrossRef]

54. Sagor, G.H.M.; Berberich, T.; Kojima, S.; Niitsu, M.; Kusano, T. Spermine modulates the expression of two probable polyamine transporter genes and determines growth responses to cadaverine in Arabidopsis. *Plant Cell Rep.* **2016**, *35*, 1247–1257. [CrossRef] [PubMed]

55. Wang, W.; Liu, J.-H. CsPAO4 of *Citrus sinensis* functions in polyamine terminal catabolism and inhibits plant growth under salt stress. *Sci. Rep.* **2016**, *6*, 31384. [CrossRef] [PubMed]

56. Bordenave, C.D.; Mendoza, C.G.; Bremont, J.F.J.; Garriz, A.; Rodriguez, A.A. Defining novel plant polyamine oxidase subfamilies through molecular modeling and sequence analysis. *BMC Evol. Biol.* **2019**, *19*, 28. [CrossRef] [PubMed]

57. Michael, A.J.; Furze, J.M.; Rhodes, M.J.; Burtin, D. Molecular cloning and functional identification of a plant ornithine decarboxylase cDNA. *Biochem. J.* **1996**, *314*, 241–248. [CrossRef] [PubMed]

58. Hanzawa, Y.; Imai, A.; Michael, A.J.; Komeda, Y.; Takahashi, T. Characterization of the spermidine synthase-related gene family in *Arabidopsis thaliana*. *FEBS Lett.* **2002**, *527*, 176–180. [CrossRef]

59. Sagor, G.H.M.; Liu, T.B.; Takahashi, H.; Niitsu, M.; Berberich, T.; Kusano, T. Longer uncommon polyamines have a stronger defense gene-induction activity and a higher suppressing activity of *Cucumber mosaic virus* multiplication compared to that of spermine in *Arabidopsis thaliana*. *Plant Cell Rep.* **2013**, *32*, 1477–1488. [CrossRef] [PubMed]

60. Sagor, G.H.M.; Kusano, T.; Berberich, T. Identification of the actual coding region for polyamine oxidase 6 from rice (OsPAO6) and its partial characterization. *Plant Signal. Behav.* **2017**, *12*, e1359456. [CrossRef] [PubMed]

61. Fincato, P.; Moschou, P.N.; Spedaletti, V.; Tavazza, R.; Angelini, R.; Federico, R.; Roubelakis-Angelakis, K.A.; Tavladoraki, P. Functional diversity inside the Arabidopsis polyamine oxidase gene family. *J. Exp. Bot.* **2011**, *62*, 1155–1168. [CrossRef] [PubMed]

62. Takahashi, T.; Kakehi, J.I. Polyamines, ubiquitous polycations with unique roles in growth and stress responses. *Ann. Bot Lond.* **2010**, *105*, 1–6. [CrossRef]

63. Hao, Y.; Huang, B.; Jia, D.; Mann, T.; Jiang, X.; Qiu, Y.; Niitsu, M.; Berberich, T.; Kusano, T.; Liu, T. Identification of seven polyamine oxidase genes in tomato (*Solanum lycopersicum* L.) and their expression profiles under physiological and various stress conditions. *J. Plant Physiol.* **2018**, *228*, 1–11. [CrossRef]

64. Kamada-Nobusada, T.; Hayashi, M.; Fukazawa, M.; Sakakibara, H.; Nishimura, M. A putative peroxisomal polyamine oxidase, AtPAO4, is involved in polyamine catabolism in *Arabidopsis thaliana*. *Plant Cell Physiol.* **2008**, *49*, 1272–1282. [CrossRef] [PubMed]

65. Fincato, P.; Moschou, P.N.; Ahou, A.; Angelini, R.; Roubelakis-Angelakis, K.A.; Federico, R.; Tavladoraki, P. The members of *Arabidopsis thaliana* PAO gene family exhibit distinct tissue and organ-specific expression pattern during seedling growth and flower development. *Amino Acids* **2012**, *42*, 831–841. [CrossRef] [PubMed]

66. Sequera-Mutiozabal, M.I.; Erban, A.; Kopka, J.; Atanasov, K.E.; Bastida, J.; Fotopoulos, V.; Alcázar, R.; Tiburcio, A.F. Global metabolic profiling of Arabidopsis polyamine oxidase 4 (AtPAO4) loss-of-function mutants exhibiting delayed dark-induced senescence. *Front. Plant Sci.* **2016**, *7*, 173. [CrossRef] [PubMed]

67. Zarza, X.; Atanasov, K.E.; Marco, F.; Arbona, V.; Carrasco, P.; Kopka, J.; Fotopoulos, V.; Munnik, T.; Gómez-Cadenas, A.; Tiburcio, A.F.; et al. Polyamine oxidase 5 loss-of-function mutations in *Arabidopsis thaliana* trigger metabolic and transcriptional reprogramming and promote salt stress tolerance. *Plant Cell Environ.* **2017**, *40*, 527–542. [CrossRef] [PubMed]

68. Alabdallah, O.; Ahou, A.; Mancuso, N.; Pompili, V.; Macone, A.; Pashkoulov, D.; Stano, P.; Cona, A.; Angelini, R.; Tavladoraki, P. The Arabidopsis polyamine oxidase/dehydrogenase 5 interferes with cytokinin and auxin signaling pathways to control xylem differentiation. *J. Exp. Bot.* **2017**, *68*, 997–1012. [CrossRef] [PubMed]

69. Ahou, A.; Martignago, D.; Alabdallah, O.; Tavazza, R.; Stano, P.; Macone, A.; Rambla, J.L.; Vera-Sirera, F.; Angelini, R.; Federico, R.; et al. A plant spermine oxidase/dehydrogenase regulated by the proteasome and polyamines. *J. Exp. Bot.* **2014**, *65*, 1585–1603. [CrossRef] [PubMed]

70. Takahashi, Y.; Ono, K.; Akamine, Y.; Asano, T.; Ezaki, M.; Mouri, I. Highly-expressed polyamine oxidases catalyze polyamine back conversion in *Brachypodium distachyon*. *J. Plant Res.* **2018**, *131*, 341–348. [CrossRef] [PubMed]

71. Wang, W.; Liu, J.H. Genome-wide identification and expression analysis of the polyamine oxidase gene family in sweet orange (*Citrus sinensis*). *Gene* **2015**, *555*, 421–429. [CrossRef]

72. Clay, N.K.; Nelson, T. Arabidopsis *thickvein* mutation affects vein thickness and organ vascularization, and resides in a provascular cell-specific spermine synthase involved in vein definition and in polar auxin transport. *Plant Physiol.* **2005**, *138*, 767–777. [CrossRef]

73. Kakehi, J.; Kuwashiro, Y.; Motose, H.; Igarashi, K.; Takahashi, T. Norspermine substitutes for thermospermine in the control of stem elongation in *Arabidopsis thaliana*. *FEBS Lett.* **2010**, *584*, 3042–3046. [CrossRef]

74. Tsaniklidis, G.; Kotsiras, A.; Tsafouros, A.; Roussos, P.A.; Aivalakis, G.; Katinakis, P.; Delis, C. Spatial and temporal distribution of genes involved in polyamine metabolism during tomato fruit development. *Plant Physiol. Biochem.* **2016**, *100*, 27–36. [CrossRef] [PubMed]

75. Guo, Z.; Tan, J.; Zhuo, C.; Wang, C.; Xiang, B.; Wang, Z. Abscisic acid, H_2O_2 and nitric oxide interactions mediated cold-induced *S*-adenosylmethionine synthetase in *Medicgo sativa* subsp. falcata that confers cold tolerance through up-regulating polyamine oxidation. *Plant Biotechnol. J.* **2014**, *12*, 601–612. [PubMed]

76. Zhuo, C.; Liang, L.; Zhao, Y.; Guo, Z.; Lu, S. A cold responsive ethylene responsive factor from *Medicago falcata* confers cold tolerance by up-regulation of polyamine turnover, antioxidant protection, and proline accumulation. *Plant Cell Environ.* **2018**, *41*, 2021–2032. [CrossRef] [PubMed]

77. Diao, Q.; Song, Y.; Shi, D.; Qi, H. Interaction of polyamines, abscisic acid, nitric oxide, and hydrogen peroxide under chilling stress in tomato (*Lycopersicon esculentum* Mill.) seedlings. *Front. Plant Sci.* **2017**, *8*, 203. [CrossRef] [PubMed]

78. Parvin, S.; Lee, O.R.; Sathiyaraj, G.; Khorolragchaa, A.; Kim, Y.J.; Yang, D.C. Spermidine alleviates the growth of saline-stressed ginseng seedlings through antioxidative defense system. *Gene* **2014**, *537*, 70–78. [CrossRef]

79. Recalde, L.; Vázquez, A.; Groppa, M.D.; Benavides, M.P. Reactive oxygen species and nitric oxide are involved in polyamine-induced growth inhibition in wheat plants. *Protoplasma* **2018**, *255*, 1295–1307. [CrossRef] [PubMed]

80. Takács, Z.; Poór, P.; Tari, I. Comparison of polyamine metabolism in tomato plants exposed to different concentrations of salicylic acid under light or dark conditions. *Plant Physiol. Biochem.* **2016**, *108*, 266–278. [CrossRef] [PubMed]

81. Ozawa, R.; Bertea, C.M.; Foti, M.; Narayana, R.; Arimura, G.; Muroi, A.; Horiuchi, J.; Nishioka, T.; Maffei, M.E.; Takabayashi, J. Exogenous polyamines elicit herbivore-induced volatiles in lima bean leaves: Involvement of calcium, H_2O_2 and jasmonic acid. *Plant Cell Physiol.* **2009**, *50*, 2183–2199. [CrossRef]

82. Wang, Y.; Ye, X.; Yang, K.; Shi, Z.; Wang, N.; Yang, L.; Chen, J. Characterization, expression, and functional analysis of polyamine oxidases and their role in selenium-induced hydrogen peroxide production in *Brassica rapa*. *J. Sci. Food Agric.* **2019**, *99*, 4082–4093. [CrossRef]

83. Dong, Q.; Magwanga, R.O.; Cai, X.; Lu, P.; Nyangasi Kirungu, J.; Zhou, Z.; Wang, X.; Wang, X.; Xu, Y.; Hou, Y.; et al. RNA-sequencing, physiological and RNAi analyses provide insights into the response mechanism of the ABC-mediated resistance to *Verticillium dahliae* infection in cotton. *Genes* **2019**, *10*, 110.

84. Jasso-Robles, F.I.; Jiménez-Bremont, J.F.; Becerra-Flora, A.; Juárez-Montiel, M.; Gonzalez, M.E.; Pieckenstain, F.L.; García de la Cruz, R.F.; Rodríguez-Kessler, M. Inhibition of polyamine oxidase activity affects tumor development during the maize-Ustilago maydis interaction. *Plant Physiol. Biochem.* **2016**, *102*, 115–124. [CrossRef] [PubMed]

85. Lulai, E.C.; Neubauer, J.D.; Olson, L.L.; Suttle, J.C. Wounding induces changes in tuber polyamine content, polyamine metabolic gene expression, and enzyme activity during closing layer formation and initiation of wound periderm formation. *J. Plant Physiol.* **2015**, *176*, 89–95. [CrossRef] [PubMed]

86. Sudhakar, C.; Veeranagamallaiah, G.; Nareshkumar, A.; Sudhakarbabu, O.; Sivakumar, M.; Pandurangaiah, M.; Kiranmai, K.; Lokesh, U. Polyamine metabolism influences antioxidant defense mechanism in foxtail millet (*Setaria italica* L.) cultivars with different salinity tolerance. *Plant Cell Rep.* **2015**, *34*, 141–156. [CrossRef] [PubMed]

87. Chen, T.; Xu, Y.; Wang, J.; Wang, Z.; Yang, J.; Zhang, J. Polyamines and ethylene interact in rice grains in response to soil drying during grain filling. *J. Exp. Bot.* **2013**, *64*, 2523–2538. [CrossRef] [PubMed]

88. Polticelli, F.; Salvi, D.; Mariottini, P.; Amendola, R.; Cervelli, M. Molecular evolution of the polyamine oxidase gene family in Metazoa. *BMC Evol. Biol.* **2012**, *12*, 90. [CrossRef] [PubMed]

89. Fu, X.Z.; Chen, C.W.; Wang, Y.; Liu, J.H.; Moriguchi, T. Ectopic expression of *MdSPDS1* in sweet orange (*Citrus sinensis* Osbeck) reduces canker susceptibility: Involvement of H_2O_2 production and transcriptional alteration. *BMC Plant Biol.* **2011**, *11*, 55. [CrossRef] [PubMed]

90. Angelini, R.; Cona, A.; Federico, R.; Fincato, P.; Tavladoraki, P.; Tisi, A. Plant amine oxidases "on the move": An update. *Plant Physiol. Biochem.* **2010**, *48*, 560–564. [CrossRef]

91. Xue, B.; Zhang, A.; Jiang, M. Involvement of polyamine oxidase in abscisic acid-induced cytosolic antioxidant defense in leaves of maize. *J. Integr. Plant Biol.* **2009**, *51*, 225–234. [CrossRef]

92. Moschou, P.N.; Sarris, P.F.; Skandalis, N.; Andriopoulou, A.H.; Paschalidis, K.A.; Panopoulos, N.J.; Roubelakis-Angelakis, K.A. Engineered polyamine catabolism preinduces tolerance of tobacco to bacteria and oomycetes. *Plant Physiol.* **2009**, *149*, 1970–1981. [CrossRef]

93. Moschou, P.N.; Delis, I.D.; Paschalidis, K.A.; Roubelakis-Angelakis, K.A. Transgenic tobacco plants overexpressing polyamine oxidase are not able to cope with oxidative burst generated by abiotic factors. *Physiol. Plant* **2008**, *133*, 140–156. [CrossRef]

94. Yoda, H.; Hiroi, Y.; Sano, H. Polyamine oxidase is one of the key elements for oxidative burst to induce programmed cell death in tobacco cultured cells. *Plant Physiol.* **2006**, *142*, 193–206. [CrossRef] [PubMed]

95. Yoda, H.; Yamaguchi, Y.; Sano, H. Induction of hypersensitive cell death by hydrogen peroxide produced through polyamine degradation in tobacco plants. *Plant Physiol.* **2003**, *132*, 1973–1981. [CrossRef] [PubMed]

96. Cona, A.; Cenci, F.; Cervelli, M.; Federico, R.; Mariottini, P.; Moreno, S.; Angelini, R. Polyamine oxidase, a hydrogen peroxide-producing enzyme, is up-regulated by light and down-regulated by auxin in the outer tissues of the maize mesocotyl. *Plant Physiol.* **2003**, *131*, 803–813. [CrossRef] [PubMed]

97. Hatmi, S.; Trotel-Aziz, P.; Villaume, S.; Couderchet, M.; Clément, C.; Aziz, A. Osmotic stress-induced polyamine oxidation mediates defence responses and reduces stress-enhanced grapevine susceptibility to *Botrytis cinerea*. *J. Exp. Bot.* **2014**, *65*, 75–88. [CrossRef] [PubMed]

98. Aloisi, I.; Cai, G.; Serafini-Fracassini, D.; Del Duca, S. Polyamines in pollen: From microsporogenesis to fertilization. *Front. Plant Sci.* **2016**, *7*, 155. [CrossRef] [PubMed]

99. Cvikrová, M.; Gemperlová, L.; Martincová, O.; Vanková, R. Effect of drought and combined drought and heat stress on polyamine metabolism in proline-over-producing tobacco plants. *Plant Physiol. Biochem.* **2013**, *73*, 7–15. [CrossRef] [PubMed]

100. Cvikrová, M.; Gemperlová, L.; Dobrá, J.; Martincová, O.; Prásil, I.T.; Gubis, J.; Vanková, R. Effect of heat stress on polyamine metabolism in proline-over-producing tobacco plants. *Plant Sci.* **2012**, *182*, 49–58. [CrossRef]

101. Filippou, P.; Antoniou, C.; Fotopoulos, V. The nitric oxide donor sodium nitroprusside regulates polyamine and proline metabolism in leaves of *Medicago truncatula* plants. *Free Radic. Biol. Med.* **2013**, *56*, 172–183. [CrossRef]

102. Xiong, H.; Guo, H.; Xie, Y.; Zhao, L.; Gu, J.; Zhao, S.; Li, J.; Liu, L. RNAseq analysis reveals pathways and candidate genes associated with salinity tolerance in a spaceflight-induced wheat mutant. *Sci. Rep.* **2017**, *7*, 2731. [CrossRef]

103. Sagor, G.H.M.; Kusano, T.; Berberich, T. Polyamine oxidase from *Selaginella lepidophylla* (SelPAO5) can replace AtPAO5 in Arabidopsis through converting thermospermine to norspermidine instead to spermidine. *Plants* **2019**, *8*, 99. [CrossRef]

104. Brikis, C.J.; Zarei, A.; Chiu, G.Z.; Deyman, K.L.; Liu, J.; Trobacher, C.P.; Hoover, G.J.; Subedi, S.; DeEll, J.R.; Bozzo, G.G.; et al. Targeted quantitative profiling of metabolites and gene transcripts associated with 4-aminobutyrate (GABA) in apple fruit stored under multiple abiotic stresses. *Hortic. Res.* **2018**, *5*, 61. [CrossRef] [PubMed]

105. Xu, X.; Shi, G.; Jia, R. Changes of polyamine levels in roots of *Sagittaria sagittifolia* L. under copper stress. *Environ. Sci. Pollut. Res. Int.* **2012**, *19*, 2973–2982. [CrossRef] [PubMed]

106. Yang, H.; Shi, G.; Wang, H.; Xu, Q. Involvement of polyamines in adaptation of *Potamogeton crispus* L. to cadmium stress. *Aquat. Toxicol.* **2010**, *100*, 282–288. [CrossRef] [PubMed]

107. Cona, A.; Moreno, S.; Cenci, F.; Federico, R.; Angelini, R. Cellular re-distribution of flavin-containing polyamine oxidase in differentiating root and mesocotyl of *Zea mays* L. seedlings. *Planta* **2005**, *221*, 265–276. [CrossRef] [PubMed]

108. Rea, G.; de Pinto, M.C.; Tavazza, R.; Biondi, S.; Gobbi, V.; Ferrante, P.; De Gara, L.; Federico, R.; Angelini, R.; Tavladoraki, P. Ectopic expression of maize polyamine oxidase and pea copper amine oxidase in the cell wall of tobacco plants. *Plant Physiol.* **2004**, *134*, 1414–1426. [CrossRef] [PubMed]

109. Gomez-Jimenez, M.C.; Paredes, M.A.; Gallardo, M.; Sanchez-Calle, I.M. Mature fruit abscission is associated with up-regulation of polyamine metabolism in the olive abscission zone. *J. Plant Physiol.* **2010**, *167*, 1432–1441. [CrossRef]

110. Paschalidis, K.A.; Roubelakis-Angelakis, K.A. Sites and regulation of polyamine catabolism in the tobacco plant. Correlations with cell division/expansion, cell cycle progression, and vascular development. *Plant Physiol.* **2005**, *138*, 2174–2184. [CrossRef] [PubMed]

111. Wu, J.; Shang, Z.; Wu, J.; Jiang, X.; Moschou, P.N.; Sun, W.; Roubelakis-Angelakis, K.A.; Zhang, S. Spermidine oxidase-derived H_2O_2 regulates pollen plasma membrane hyperpolarization-activated Ca^{2+}-permeable channels and pollen tube growth. *Plant J.* **2010**, *63*, 1042–1053. [CrossRef]

112. Agudelo-Romero, P.; Bortolloti, C.; Pais, M.S.; Tiburcio, A.F.; Fortes, A.M. Study of polyamines during grape ripening indicate an important role of polyamine catabolism. *Plant Physiol. Biochem.* **2013**, *67*, 105–119. [CrossRef]

113. Planas-Portell, J.; Gallart, M.; Tiburcio, A.F.; Altabella, T. Copper-containing amine oxidases contribute to terminal polyamine oxidation in peroxisomes and apoplast of *Arabidopsis thaliana*. *BMC Plant Biol.* **2013**, *13*, 109. [CrossRef]

 *plants*

Article

Genetically Modified Heat Shock Protein90s and Polyamine Oxidases in *Arabidopsis* Reveal Their Interaction under Heat Stress Affecting Polyamine Acetylation, Oxidation and Homeostasis of Reactive Oxygen Species

Imene Toumi [1,*], Marianthi G. Pagoulatou [1], Theoni Margaritopoulou [2,†], Dimitra Milioni [2] and Kalliopi A. Roubelakis-Angelakis [1,*]

[1]   Department of Biology, University of Crete, Voutes University Campus, 70013 Heraklion, Greece
[2]   Department of Plant Biotechnology, Agricultural University of Athens, Iera odos 75, 11855 Athens, Greece
*   Correspondence: imtoumdem@gmail.com (I.T.); poproube@uoc.gr (K.A.R.-A.)
†   Present Address: Benaki Phytopathological Institute, Department of Phytopathology, Laboratory of Mycology, St. Delta 8, 14561 Kifissia, Greece.

Received: 22 June 2019; Accepted: 12 August 2019; Published: 3 September 2019

Abstract: The chaperones, heat shock proteins (HSPs), stabilize proteins to minimize proteotoxic stress, especially during heat stress (HS) and polyamine (PA) oxidases (PAOs) participate in the modulation of the cellular homeostasis of PAs and reactive oxygen species (ROS). An interesting interaction of HSP90s and PAOs was revealed in *Arabidopsis thaliana* by using the *pLFY:HSP90RNAi* line against the four *AtHSP90* genes encoding cytosolic proteins, the T-DNA *Athsp90-1* and *Athsp90-4* insertional mutants, the *Atpao3* mutant and pharmacological inhibitors of HSP90s and PAOs. Silencing of all cytosolic *HSP90* genes resulted in several-fold higher levels of soluble spermidine (S-Spd), acetylated Spd (N^8-acetyl-Spd) and acetylated spermine (N^1-acetyl-Spm) in the transgenic *Arabidopsis thaliana* leaves. Heat shock induced increase of soluble-PAs (S-PAs) and soluble hydrolyzed-PAs (SH-PAs), especially of SH-Spm, and more importantly of acetylated Spd and Spm. The silencing of *HSP90* genes or pharmacological inhibition of the HSP90 proteins by the specific inhibitor radicicol, under HS stimulatory conditions, resulted in a further increase of PA titers, N^8-acetyl-Spd and N^1-acetyl-Spm, and also stimulated the expression of PAO genes. The increased PA titers and PAO enzymatic activity resulted in a profound increase of PAO-derived hydrogen peroxide (H_2O_2) levels, which was terminated by the addition of the PAO-specific inhibitor guazatine. Interestingly, the loss-of-function *Atpao3* mutant exhibited increased mRNA levels of selected *AtHSP90* genes. Taken together, the results herein reveal a novel function of HSP90 and suggest that HSP90s and PAOs cross-talk to orchestrate PA acetylation, oxidation, and PA/H_2O_2 homeostasis.

Keywords: heat shock proteins; heat stress; polyamines; polyamine oxidases; PA acetylation; PA oxidation; PA back-conversion; hydrogen peroxide

---

1. Introduction

Heat stress impairs plant growth and productivity globally [1]. Plants, as sessile organisms, are highly adaptive to harsh environmental conditions and accommodate the developmental programming in a wide temperature range. Survival above optimal temperature conditions is accompanied by a massive accumulation of HSPs [2–4]. HSPs maintain and stabilize proteins and transcription factors through heterocomplex formation and folding/unfolding, thus controlling the activation/inhibition of their complex associates [5–8]. Furthermore, HSPs bind to targeted molecules through specific receptors for translocation to non-peptide target molecules, such as hormones [9,10], and they act in all cellular compartments [11].

In *Arabidopsis thaliana*, the *HSP90* gene family consists of seven members [12–16]. Four highly homologous *HSP90* genes (*AtHSP90-1, -2, -3* and *-4*) encode cytosolic proteins, suggesting important functional redundancies [17,18]. They associate with key components of essential cellular processes [19] and play crucial role(s) in areas as diverse as cellular homeostasis, cell growth, development, and organismal evolution [5,20–22]. Recent evidence has highlighted the commitment of HSP90 proteins on plant defense mechanisms [23], brassinosteroid or auxin signaling pathways [10,24,25], and stomata patterning [26]. Moreover, HSP90s are essential for the vegetative-to-reproductive phase transition and flower development [22].

The inhibition of HSP90 increases the stochastic variation inherent to developmental processes by releasing morphogenic traits controlled by the HSP90 buffering capacity of mutations. Administration of radicicol (specific HSP90 inhibitor) to wild type (WT) plants reduces viability under physiological conditions [5], whereas many homozygous *hsp90* mutant lines are lethal [15,27]. In *Arabidopsis*, the *Athsp90c-1* mutant exhibits a high rate of albinos and aborted seeds, confirming that the HSP90 protein localized specifically to chloroplasts is essential for viability [27]. HSP90s interact with tetrapyrroles, by modulating the photosynthesis-associated nuclear genes (*PhANG*) expression, in response to oxidative stress [28]. They associate with the suppressor of the G2 allele of *skp1* (*SGT1*) for the stabilization of the nucleotide-binding domain and leucine-rich repeat-containing (NLR) immune sensors, which mediate plant defense mechanisms [29]. In addition, HSP90s participate in stress signaling pathways [18,30–33]. In fact, under stress, the inhibition of HSP90 induces, in most cases, deregulation of the putative client's functionality and alteration of cell processes [31,34].

Polyamines are highly reactive aliphatic polycations [35]. The more abundant and best studied PAs are the diamine putrescine (Put), the triamine Spd and the tetramine Spm. Polyamine homeostasis is determined by a complex regulatory mechanism which includes turnover, export/transport, as well as modifications of their amino-groups, *via* mechanisms such as acetylation and conjugation [36–39]. Polyamines, *per se* or *via* their metabolic products such as H_2O_2, interfere with a plethora of dynamic metabolic and developmental processes [40–42], as well as stress responses [37,38,43–47]. Also, PAs are highly involved in adaptive mechanisms of hyperthermophilic proteins [48], whereas they interfere with HSP synthesis under increasing temperatures by an unknown mechanism [49].

Polyamine oxidases participate in the regulation of PA homeostasis, mediating their oxidation/ back-conversion [39,50]. In *Arabidopsis thaliana*, oxidation/back-conversion of PAs is mediated by the peroxisomal *AtPAO2* (At2g43020), *AtPAO3* (At3g59050), *AtPAO4* (At1g65840) and the cytosolic *AtPAO1* (At5g13700) and *AtPAO5* (At4g29720) [46]. Products of PAO-mediated enzymatic action include aldehydes and H_2O_2 [43,47]. Recently, PAOs were confirmed to play an important role in the control of cell proliferation (in animals) through the generated toxic aldehydes and H_2O_2 and were proposed as fine competitors for antiproliferative therapies [51]. In plants, the generated H_2O_2, depending on its "signature", can signal either the orchestration of tolerance to abiotic/biotic stresses or execution of programmed cell death (PCD) [37–39,52–60]. Furthermore, an NADPH-oxidase/PAO feedback loop controls oxidative burst under salinity in tobacco [61]. In animals [62], but not in plants [47,63], acetylated PAs are the major form of the conjugated PAs and the favorite substrate for PAOs.

Evidence for the interplay between HSPs and PAs was first provided for the HSP70 family. Depletion of PAs in a cell line derived from rat hepatoma induces direct inhibition of *HSP70* expression [64]. In addition, exogenous Spm increases the transcription of *HSP90* and subsequently enhances heat tolerance in *Arabidopsis thaliana* [65]. Furthermore, the underexpression of PAO results in thermotolerant tobacco plants [66]. Such findings suggest the existence of an evolutionary conserved network which acts for co-regulation of PAs and HSPs to efficiently orchestrate HS response. However, the exact mechanism remains elusive. In this work, we attempted to identify potential links between *HSP90* and PA homeostasis implicating more specifically *PAO* genes, PAO activity and H_2O_2 levels. Impaired *HSP90* action, derived either from the *HSP90RNAi* effect or from pharmacological inhibition of the HSP90 protein activity, induced increase in PA titers, especially of the acetylated-PAs, N^8-acetyl-Spd and N^1-acetyl-Spm and the conjugated forms. The interconnection between HSP90 and PAs was further determined since loss of PA homeostasis by impairing a major path of PA oxidation had a specific impact on *HSP90* both at transcriptional and translational levels. Similarly, genetic modification of HSP90 had a profound effect on genes involved in the PA oxidation pathway. Our results reveal a novel interaction by which *HSP90s/PAOs* co-regulate the intracellular titers of PAs/H_2O_2 with an unknown yet impact on stress response.

2. Results

2.1. Underexpression of AtHSP90 1–4 Genes Results in Higher S-PAs and S-N^8/N^1 Acetylated PA Titers

Polyamine homeostasis and *PAO* expression/PAO activity are cell/tissue/organ- specific, strongly affected by the ontogenetic stage of the plant/organ and the growing conditions [40,41]. In order to identify any potential link between HSP90 and PA homeostasis, the endogenous PA titers were analyzed in the leaves of 15 day-old WT, *Athsp90-1* and *Athsp90-4* mutants and *pLFY:HSP90RNAi* transgenic plants. More specifically, the titers of soluble Put (S-Put), S-Spd, S-Spm, and the two mono-acetylated PAs, N^8-acetyl-Spd and N^1-acetyl-Spm were determined (Figure 1A). The leaves of *Athsp90 1–4* contained significantly higher levels of S-Spm and acetyl-PAs and the *pLFY:HSP90RNAi* line contained higher levels of S-Put, S-Spd and S-Spm compared to WT. Interestingly, the *pLFY:HSP90RNAi* transgenics exhibited a nearly 18-fold increase of (N^8-acetyl-Spd + N^1-acetyl-Spm) compared to the WT (Figure 1A).

To further confirm that genetic inhibition of the *HSP90* genes affects PA homeostasis independently of plant age, PA contents were determined in 35 day-old mutants/transgenics and WT plants. Consistently, although a general decrease in PA levels was observed compared to the 15 day-old seedlings, in accordance with our previous results for tobacco and grapevine [40], all three lines, *Athsp90 1–4*, and *pLFY:HSP90RNAi* contained significantly greater S-Spm with the highest levels recorded in the *pLFY:HSP90RNAi* line, which also contained higher S-Put (Figure 1B). The most striking effect of *AtHSP90* deregulation on PA titers was, as in the 15 day-old plants, the significant increase of S-acetylated forms of Spd (N^8-acetyl-Spd) and Spm (N^1-acetyl-Spm) in the *Athsp90* transgenic lines (Figure 1B). The picture was somehow different in the conjugated SH-fraction (Figure S1). The *Athsp90 1–4* lines contained significantly higher SH-Put, SH-Spd and SH-Spm, as well as SH-acetyl-PAs, when compared to the corresponding titers in the S-fraction whereas the *pLFY:HSP90RNAi* line contained significantly lower amounts of SH- and acetyl-PAs, when compared to the *Athsp90 1–4* mutants and to WT (Figure S1). The results point strongly to a complex mechanism connecting *HSP90* and PA homeostasis which may be ontogenetically regulated.

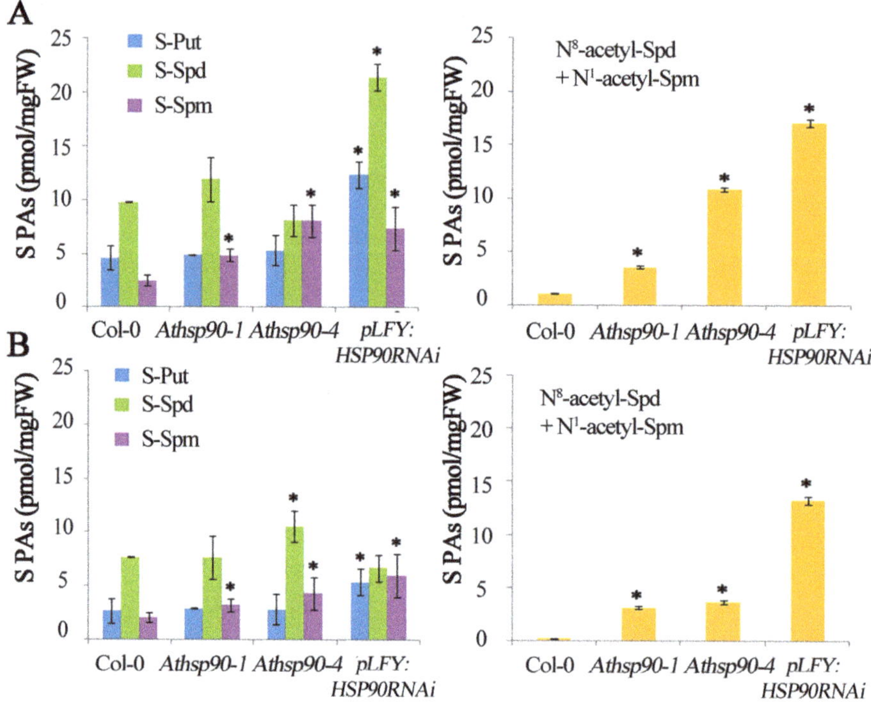

Figure 1. Endogenous polyamine titers in WT, *Athsp90-1*, *Athsp90-4* and *pLFY:HSP90RNAi* mutants of *Arabidopsis thaliana*. (**A**) Soluble Putrescine, Spermidine and Spermine (S-Put, S-Spd, S-Spm) and Soluble acetylated (N^8-acetyl-Spd + N^1-acetyl-Spm) polyamine contents in leaves of 15 day-old seedlings. (**B**) Soluble (S-Put, S-Spd, S-Spm) and soluble acetylated (N^8-acetyl-Spd + N^1-acetyl-Spm) polyamine contents in leaves of 35 day-old WT, *Athsp90-1*, *Athsp90-4* and *pLFY:HSP90RNAi* plants. Values from 8 replicates from two independent experiments were pooled for statistical analysis. Asterisks indicate statistically significant differences from WT control ($p < 0.05$).

2.2. The Increased PAs in the Athsp90 Mutants Correlate with Increased H_2O_2 Content. HSP90s Modulate Free Radical Production

Under normal conditions, PA back-conversion catalyzed by PAOs leads to H_2O_2 production balancing the intracellular oxygen consumption rate [38]. In order to examine whether HSP90s modulate H_2O_2 production, the in situ semi-quantitative method based on staining with 3,3′-diaminobenzidine (DAB) was employed. We found that all 15 day-old *Athsp90* and *pLFY:HSP90RNAi* seedlings had increased H_2O_2 content compared to WT seedlings, as assessed by DAB staining (Figure 2). Hydrogen peroxide levels paralleled the increased S-PAs titers, which are substrates of PAOs; the *pLFY:HSP90RNAi* line as well as the *Athsp90-1* and the *Athsp90-4* young seedlings, which contained the highest S- and acetyl-PAs, exhibited the highest H_2O_2 levels compared to WT (Figures 1A and 2). These results further point to a certain correlation between *HSP90s*, PAs, and PAO-generated H_2O_2.

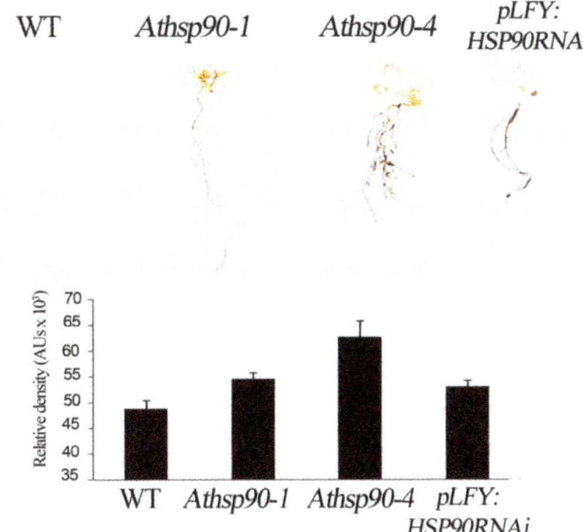

Figure 2. In situ DAB detection of hydrogen peroxide in WT, *Athsp90-1*, *Athsp90-4* and *pLFY:HSP90RNAi* seedlings of *Arabidopsis thaliana* grown on Murashige and Skoog (MS) plates and relative pixels density of leaf staining. Values from 8 replicates from two independent experiments were pooled for statistical analysis.

2.3. Under HS, Genetic Depletion and Pharmacological Inhibition of HSP90 Affect PA Homeostasis and Enhance Acetylated Forms

Since HS induces transcription of the *HSP90* genes [12,13,16], it was of interest to assess whether genetic depletion of *HSP90 1–4* genes by RNAi [22] or pharmacological inhibition of HSP90 in WT *Arabidopsis* plants would similarly affect PA homeostasis under HS conditions. Thus, analysis of PA titers was performed in leaves of WT and *pLFY:HSP90RNAi* transgenic plants following acute HS (1 h at 42 °C) in the presence (in WT) or absence (in *pLFY:HSP90RNAi*) of radicicol (Rad), a specific inhibitor of HSP90 activity. In WT, HS resulted in increased S-Spm, and considerable decrease of SH-Spd and SH-Spm levels (Figure 3). Soluble-Spm is a common thermo-responsive indicator and has been proposed as a marker of thermotolerance [65]. The combined effect of HS + Rad resulted in increased S-Put, S-Spd, SH-Put, SH-Spd and SH-Spm when compared to HS treatment alone. In the WT, the S- N^8-acetyl-Spd and N^1-acetyl-Spm titers were slightly affected by HS, whereas the inhibition of HSP90 by Rad resulted in significant increase of the N-acetyl-PAs in both fractions S and SH (Figure 3). Similarly, genetic depletion of cytosolic HSP90s in the *pLFY:HSP90RNAi* line resulted in a significant increase of all S-PAs, SH-PAs as well as of S-acetyl-PAs and SH-acetyl-PAs under HS. In particular, S-Spd in *pLFY:HSP90RNAi* showed a nearly 3-fold increase (Figure 3). These data indicate that impairment of the HSP90 molecular chaperones, when coupled to the stimulatory effect of the HS-related pathway, increases further the deregulation of PA homeostasis and leads to a significant increase of acetylated PAs.

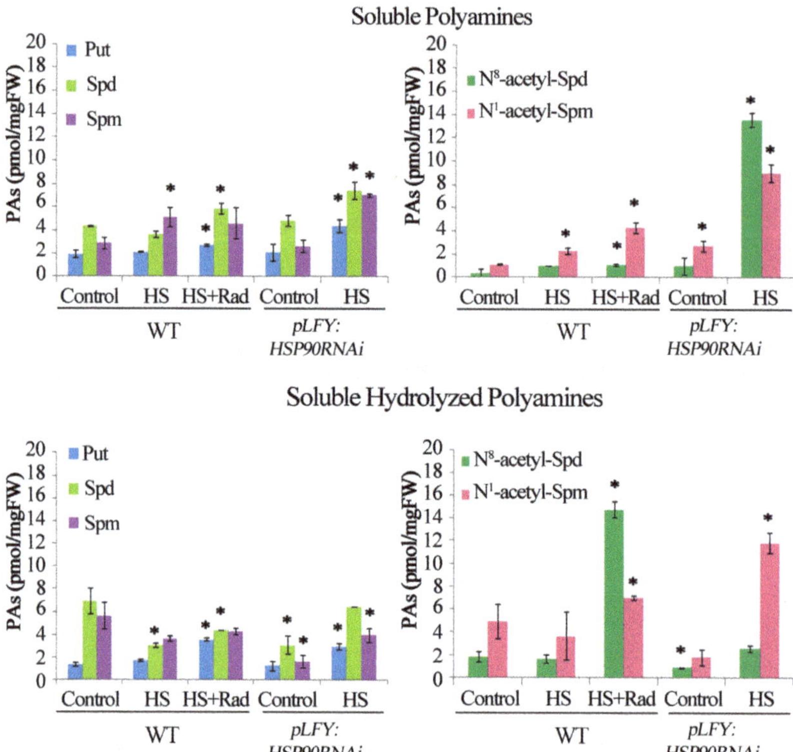

Figure 3. HSP90 activation/inhibition affects polyamine homeostasis in *Arabidopsis thaliana*. Endogenous Soluble (Put, Spd, Spm); Soluble acetylated (N^8-acetyl-Spd and N^1-acetyl-Spm); Soluble hydrolyzed (Put, Spd, Spm); and Soluble hydrolyzed acetylated (N^8-acetyl-Spd and N^1-acetyl-Spm) polyamine contents following HSP90 activation/inhibition in treated leaves. Leaves of WT and *pLFY:HSP90RNAi* transgenic plants followed brief HS (1 h at 42 °C) in the presence (in WT) or absence (in *pLFY:HSP90RNAi*) of radicicol (Rad). Values from 8 replicates from two independent experiments were pooled for statistical analysis. Asterisks indicate statistically significant differences from the WT control ($p < 0.05$).

2.4. Genetic and Pharmacological Inhibitions of HSP90 Affect the PA Oxidation Pathway

Taking into consideration the changes in PA titers and the high levels of H_2O_2 in the RNAi *pLFY:HSP90RNAi* line (Figures 1 and 2), we studied the expression profile of *AtPAO1, 3, 5* and *AtHSP90 1–4* genes in WT and this transgenic line. To explore the impact that *HSP90* could have on PA homeostasis, we studied the expression of *AtPAO1, AtPAO3* and *AtPAO5* genes in *pLFY:HSP90RNAi* plants in which cytosolic HSP90 proteins are markedly depleted [16,20]. In WT plants, HS increased mRNA levels of *AtHSP90-1* and *AtHSP90-4*; treatment with HS + Rad also induced high *AtHSP90-1* and *AtHSP90-4* transcript levels when compared to the controls. It is worth noticing that the abundance of *AtHSP90-1* is slightly lower when HS coupled with Rad than HS alone (Figure 4A).

AtPAO mRNAs in the *pLFY:HSP90RNAi* lines were increased when compared to WT under control conditions, which was not remarkably different under HS (Figure 4B). If the expression of *PAO* genes was affected by HSP90 function, then Rad should modulate gene transcription pattern similarly to genetically compromised HSP90 activity. In fact, Rad application on WT plants induced the expression of the *AtPAO1, AtPAO3* and *AtPAO5* genes (Figure 4B). Consequently, pharmacological HSP90 inactivation affected the expression of *PAO* genes in a similar way to *HSP90* mRNA depletion by RNAi silencing. Similar results were obtained when the abundance of PAO3 protein was assessed.

The levels of PAO3 protein correlated well with the corresponding mRNA levels in both the WT and the *pLFY:HSP90RNAi* plants under the different treatments (Figure 4C).

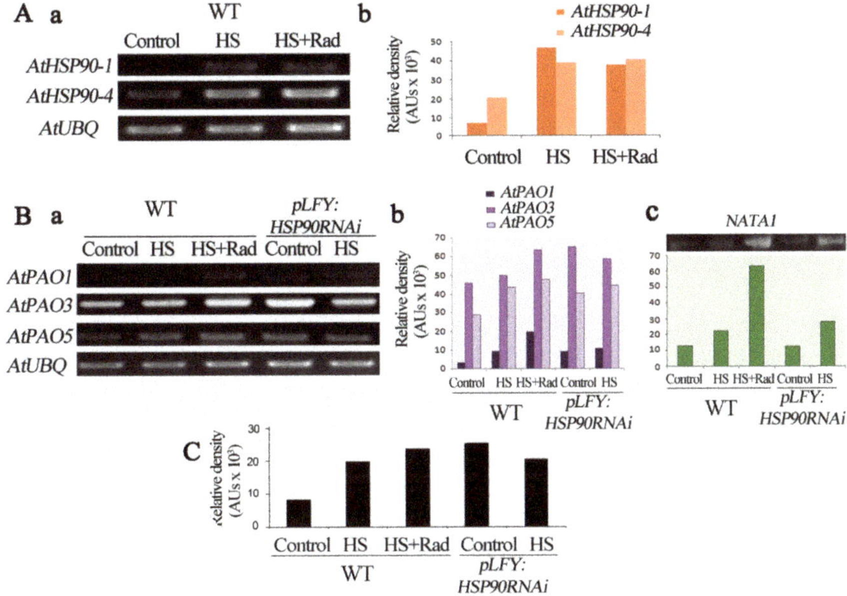

Figure 4. Inhibition of HSP90 affects levels of mRNAs of *AtPAO1*, *AtPAO3*, *AtPAO5* and *NATA1* genes, and PAO3 protein in *Arabidopsis thaliana*. (**A**) a. Abundance of transcripts of *AtHSP90-1* and *AtHSP90-4* genes in WT, b. relative pixels density of gel bands' intensities following normalization with *AtUBQ*. (**B**) a. Abundance of transcripts of *AtPAO1*, *AtPAO3* and *AtPAO5*, b. relative pixels density of gel bands' intensities following normalization with *AtUBQ*, c. *NATA1* mRNA levels in WT and *pLFY:HSP90RNAi* mutant, with relative densitometric measurements following normalization with *AtUBQ*. (**C**) PAO3 immunoreactive protein relative levels in the leaves under control, HS and HS + Rad conditions. Image software was used for numeric determination and quantification of gels bands' intensities.

2.5. Inhibition of HSP90 under HS Induces Expression of NATA1, a Putative Acetyltransferase-Like Gene

As shown above, leaves of *pLFY:HSP90RNAi* plants exposed to HS as well as leaves of WT exposed to HS coupled with Rad, exhibit a noticeable increase of acetylated forms of higher PAs (Figure 3). The *Arabidopsis NATA* genes are the closest homologues of mammalian PA acetyltransferases. Thus, due to the lack of identified genes regulating the acetylation of Spd and Spm in *Arabidopsis*, we prompted to examine the transcript levels of *NATA1* gene (At2g39030) which is identified as a Put and ornithine acetyltransferase [67,68], as a means to assess the genetic trend of PA acetylation under our conditions. Again, and by using excised leaves incubated in Morpholino Ethane Sulfonic acid (MES) and exposed to HS with (WT)/without Rad (*Athsp90s*), the WT and the *PLFY:HSP90RNAi* mutant under control conditions contained low amounts of *NATA1* mRNA. Heat shock induced *NATA1* expression in the WT and *PLFY:HSP90RNAi* plants, exhibiting a 2-fold increase of the transcript levels. Interestingly, in WT plants *NATA1* transcripts increased more than 5-fold under HS in the presence of Rad. The results show that inhibition of HSP90 by the HSP inhibitor Rad releases PA acetylation process and suggest that PA acetylation is in fact inhibited by HSP90 (Figure 4B,C).

2.6. PAO-Induced Stimulation during HSP90 Inhibition Is Responsible for the Increased H_2O_2

Since inhibition of HSP90 induced *PAO* transcription and increased PAO protein levels (Figure 4B,C), supporting cross-talk between the inhibition of HSP90 and the stimulation of the PAO pathway, we were prompted to further establish this relationship; thus, PAO activity was determined, as H_2O_2 production, in leaves of WT plants treated with HSP90 inhibitor and in *PLFY:HSP90RNAi*, under HS. The results confirmed the activation of PAO under such conditions (Figure 5A and Figure S2A). To further test this finding, we also determined PAO activity *in planta* intact WT leaves infiltrated with Rad or Spd (as a positive control) and submitted to HS (Figure 5B). In addition, the results were confirmed by determining the endogenous H_2O_2 of leaves treated as above, which confirmed that in fact high PAO activity correlates with high H_2O_2 levels (Figure 5C). Finally, the increased amounts of H_2O_2 were further verified by in situ staining of WT leaves under HS and HS + Rad. Consistently, guazatine (specific PAO inhibitor), when administered simultaneously with Rad, reduced H_2O_2 levels, supporting that the higher PAO activity contributes to H_2O_2 generation upon HSP90 inhibition (Figure 5D). As with WT, the use of guazatine along with HS attenuated PAO activity in *PLFY:HSP90RNAi* leaves (Figure S3A,B).

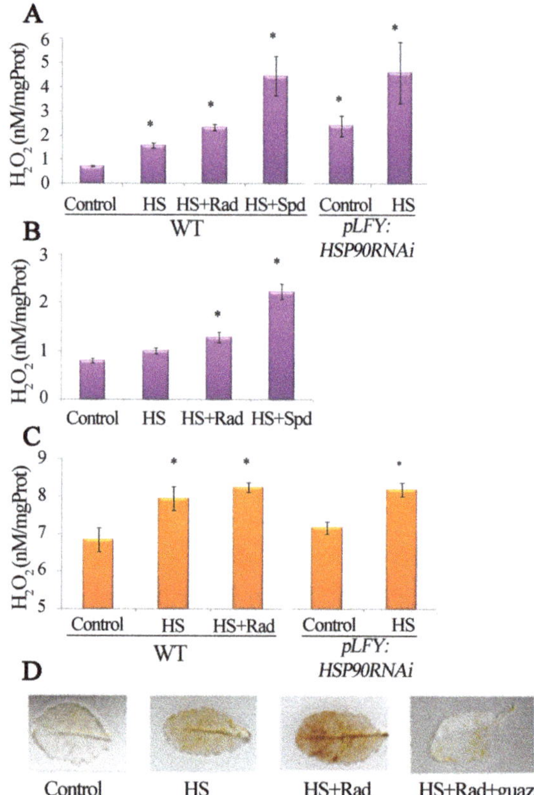

Figure 5. Inhibition of HSP90 stimulates the polyamine oxidation activity. Polyamine Oxidase (PAO) activity measured and evaluated through hydrogen peroxide production in: (**A**) Excised WT leaves cultured on MES medium and exposed to HS with/without Rad; and excised *pLFY:HSP90RNAi* line leaves cultured on MES medium and exposed to HS. (**B**) *In planta* WT leaves infiltrated separately with water (control or HS); Rad, and Spd (as positive control) prior to exposure to HS. (**C**) Luminometric hydrogen peroxide quantification in WT leaves exposed to HS and HS + Rad, and *pLFY:HSP90RNAi* mutant exposed to HS. (**D**) In situ DAB staining of hydrogen peroxide in WT control leaves treated with HS and HS + Rad and with HS + Rad + guazatine (Guaz). Values from 8 replicates from two independent experiments were pooled for statistical analysis. Asterisks indicate statistically significant differences from WT control ($p < 0.05$).

2.7. PAO and AtHSP90 Reciprocally Affect Each Other's Expression

To examine whether PAO homeostasis has an impact on *AtHSP90* expression, *Atpao3* loss-of-function plants were tested. By applying HS to leaves incubated in MES medium, transcript analysis revealed that only *AtHSP90-1* and *AtHSP90-4* mRNAs were increased in the *Atpao3* mutant compared to WT. The abundance of *AtHSP90 1–4* mRNAs as well as HSP90 protein were analyzed in the *Atpao3* mutant and the WT, under HS (Figure 6A,B). In the *Atpao3*, mRNA titers of the *AtHSP90-1* and *AtHSP90-4* were significantly greater compared to WT, whilst HS resulted in an obvious increase of the mRNA levels of the *AtHSP90 1–4* genes (Figure 6A). Inhibition of HSP90 activity simultaneously to HS stimulation (HS + Rad) exerted no further significant effect on the *AtHSP90 1–4* mRNA levels (Figure 6A).

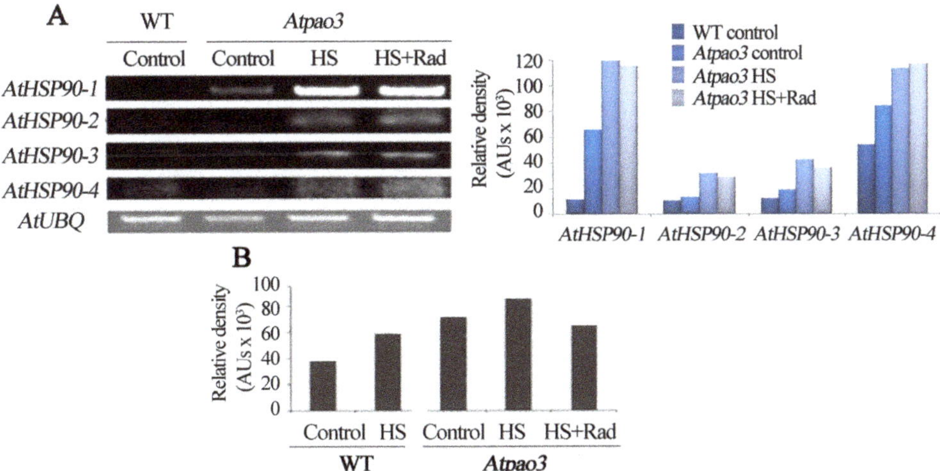

Figure 6. Abundance of cytosolic *AtHSP90* mRNA in WT and *Atpao3* mutant and HSP90 protein levels. Leaves were treated with HS and HS + Rad for 1 h at 42 °C. (**A**) Reverse transcriptase (RT)-PCR of the four *AtHSP90* cytosolic genes (*AtHSP90 1–4*) and relative intensities of bands normalized against ubiquitin. (**B**) Quantification of the HSP90 immunoreactive protein in control and HS- treated WT leaves, as well as in controls, HS and HS + Rad treated *Atpao3* leaves.

It is known that HSP90 are essential molecular chaperones in eukaryotic cells, with key functions in signal transduction networks. To test whether HSP90 homeostasis is disrupted when the PA oxidation pathway is challenged, the molecular chaperone protein abundance was assessed. Immunoreactive HSP90 protein levels exhibited parallel increase under HS in both, WT and *Atpao3* and was more abundant in the *Atpao3* genotype (Figure 6B and Figure S2B). In *pao3* mutant, the HSP90 protein levels were increased.

Furthermore, semi-quantitative reverse transcriptase (RT)-PCR revealed that the cytosolic *AtHSP90 1–4* genes in *Atpao3* seedlings (5, 10, and 15 day-old) revealed a general over-expression of these genes in the *Atpao3* mutant compared to WT, at different time points (Figure S4).

Interestingly, the *Atpao3* leaves contained lower endogenous mRNA levels of all four *AtPAO* genes (*AtPAO1, 2, 4* and *5*). Pharmacological inhibition of HSP90 activity under HS (HS + Rad) induced an increase in the *AtPAO* transcripts and restored the *AtPAO* mRNAs to levels similar to the WT control, except the *AtPAO4* that remained low (Figure S5A). The restoration of the *AtPAO* mRNA levels in the *Atpao3* mutant subsequently to the inhibitory effect of Rad on HSP90 activity suggests that *AtPAOs* are activated subsequently to *HSP90* deregulation. Application of HS + Rad did not induce accumulation of acetyl-S-PAs in the *Atpao3* mutant genotype (Figure S5B).

3. Discussion

The hypothesis of cross-talk between HSPs and PAs was initially evoked through the finding that exogenous Spm increases mRNA levels of *HSP90* and subsequently enhances heat tolerance in *Arabidopsis* [65]. Results herein support this hypothesis and reveal reciprocal molecular and biochemical interactions of *HSP90/PAO* genes and/or HSP90/PAO proteins. Depletion of cytoplasmic HSP90 results in increased levels of PAs and more specifically of N^8-acetyl-Spd and N^1-acetyl-Spm, which, under physiological conditions, are present only in traces in the leaves of WT (Figure 1) [69]. In animals, Spd/Spm acetyltransferase (SSAT) is a key enzyme participating in the first step of PA catabolism, mediating the acetylation of PAs whereas PAOs catalyze the final oxidation of acetyl-PAs. Overexpression of *SSAT* correlates with growth inhibition in a wide range of tumors [70–72]. Accumulation of acetyl-PAs by the increased activity of SSAT shifts the metabolism of PAs into lipids and carbohydrates through the increasing consumption of acetyl-CoA and ATP, leading to generation of ROS and, subsequently, to oxidative stress and proteotoxicity [71]. On the contrary, the acetyl-PAs are not the preferred substrates for PAOs in plants [39,55]. Acetylated-Put is directly linked to the pathogen-defense related signals [67] as well as the root cell de-differentiation process, probably as transient metabolite for GABA production [73]. The factors controlling the tri- and tetraamine (Spd and Spm) acetylation remain obscure. *NATA1* is one of the few genes participating in the N-acetylation of amine metabolism in *Arabidopsis*; ornithine is the favorite substrate when stimulated *via* methyl jasmonate [67] and Put as specific substrate when stimulated through jasmonate and salicylic acid [68], both involving the pathogen infection process. Analysis of the mRNA levels of the *NATA1* in *pLFY:HSP90RNAi* HS-treated leaves versus WT treated with HS +/- Rad confirms the stimulation of the *NATA1* transcription under HS in both genotypes, and a further stimulation when HS is associated with HSP90 inhibitor in the WT (Figure 4B).

Both, HSP90s and PAs, are highly responsive to environmental challenges. HSP90s increase under stress, including heat, osmotic, and heavy metals, as well as biotic stress, whereas PAs exhibit complex profiles under such conditions [1,12,32,37,38,74–76]. Spermine plays a significant role in *Arabidopsis* thermotolerance, and exogenously applied Spm enhances HS response by inducing increase of mRNA levels of *HSP* genes [65]. Although sufficient information on the implication of HSP in proteotoxicity is lacking, it is well established that HSP chaperones are highly responsible for the prevention of protein degradation and misfolding and for preventing non-native proteins from non-specific intermolecular interactions, eliminating proteotoxicity propagation [77,78]. Nishizawa-Yokoi et al. [79] propose a model for HSP90s role in ROS-induced proteotoxic stress in *Arabidopsis thaliana*. They suggest that ROS accumulation results in inhibition of proteasome 26S, which induces a peculiar poly-ubiquitination. HSP90s, which act as Heat Shock Factor (HSF) repressors under normal conditions, release the HSF which regulates in turn the expression of a Heat Shock Transcription Factor A2 (*HSFA2*). Also, HSP90s regulate the production of superoxides mediated by NADPH-oxidase but are not required for the generation of H_2O_2 in human embryonic kidney cells [75]. Pharmacological inhibition of HSP90 in PC-12 cells induces cytotoxicity and cell death *via* oxidative stress by an unidentified mechanism [34]. Taking into consideration the involvement of HSP90 in cell proliferation processes, we can speculate that both, HSP90s and PAs may be integrated in a specific biochemical pathway through the competitive regulation of cell cytotoxicity and ROS generation involving potentially the homeostatic regulation of oxidation and/or acetylation of PAs.

The drastic reduction of HSP90s results in deregulation of PA titers and induction of the PAO pathway, which generates H_2O_2. In WT, the pharmacological inhibition of HSP90s under HS induces expression of *AtPAO* genes, suggesting that HSP90s may be implicated in a way to decelerate the action of PAOs into the cell. Indeed, the *pLFY:HSP90RNAi* line exhibits elevated endogenous *AtPAO* mRNA levels similar to those in WT when treated with HS + Rad, suggesting that biochemical inhibition of HSP90s defines a favorable endogenous potential of PA oxidation (Figures 4 and 5).

As key agents in the regulation of thermotolerance, HSP90s show strong connection with other stress-related factors in signal perception and transduction [18,80]. Within this context, HS induces in both WT and *pLFY:HSP90RNAi* mutant an increase of H_2O_2, which is a predictable trait of stress (Figure 5C). Simultaneous application of HS + Rad enhances H_2O_2 generation in WT. In parallel with the induction of PA oxidation, the use of guazatine, a specific inhibitor of PAO, attenuates the HS + Rad-dependent accumulation of H_2O_2, confirming the concomitant link between HSP90 inhibition/PA oxidation-back-conversion/H_2O_2 generation. HSP90 inhibition results in *AtPAOs* expression and increased transcript levels, especially the *AtPAO1* and *AtPAO5*. Within this context and the specific affinity of the two proteins AtPAO1 and AtPAO5 to Spm (Nor and especially T-Spm) [52,81], we stress the specific implication of Spm oxidation in HS response. *AtPAO3* is also highly affected by the deletion of the HSP90 action but seems to follow different pattern. Furthermore, the transcriptional analysis of the four cytosolic *AtHSP90* genes in the *Atpao3* mutant reveals the high endogenous *AtHSP90* transcript levels in this mutant and once again the putative mutual inhibitory effect.

Overall, our results suggest a link between HSP90s and PAs/PAOs/H_2O_2 under HS. As schematically presented in the proposed model (Figure 7), HSP90 proteins participate in the regulation of PA conjugation, more specifically, in PA acetylation. The *pLFY:HSP90RNAi* plants exhibit: (i) overexpression of the major *AtPAO* genes and consequently increased PAO protein and activity levels, which result in increased H_2O_2 intrinsic levels; and (ii) significant increase of PAs and, especially, of acetylated higher PAs titers. In turn, underexpression of the *AtPAO3* gene (*Atpao3* mutant), induces overexpression of one or more of the *AtHSP90 1–4* genes. Under conditions of HS as pivotal HSP90 stimulator: (1) the transgenic reduction of the *HSP90* genes (*pLFY:HSP90RNAi*), as well as the pharmacological inhibition of the HSP90 proteins *via* Rad affect similarly both pathways. Namely: (i) positive regulation of the *AtPAOs* and further increase of the PAOs-generated H_2O_2, which is also observed under HS conditions in the WT, but to a lesser extent, suggesting that HSP90s are involved in the regulation of the stress-induced PA-oxidation/back-conversion pathway; and, (ii) increase of PAs and acetylated-higher PAs, showing a specific heat-related HSP90 pathway as revealed by HSP90 inhibition. Consequently, by being powerful modulators of PAOs, HSP90s affect the PAO-generated H_2O_2, which participates in the stress-induced downstream signaling cascade that is tidily related to ROS homeostasis and PCD [38,43,50,51,61,82,83].

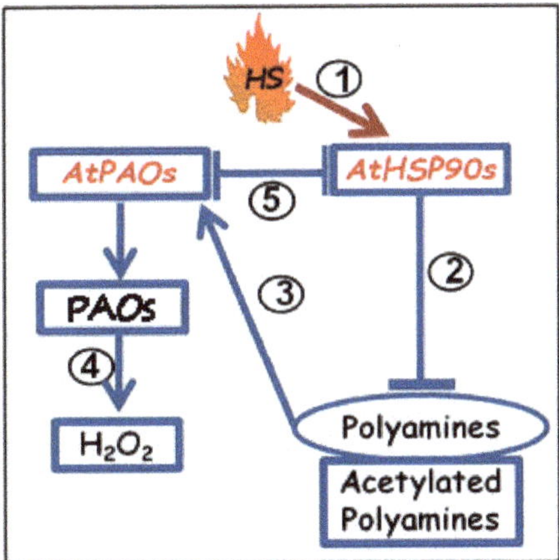

Figure 7. Proposed model of Heat Shock Protein90s (HSP90s)/Polyamine oxidases (PAOs)/Polyamines (PAs)/H_2O_2 cross-talk in *Arabidopsis thaliana*: Under conditions of HS, as pivotal HSP stimulator (1), the transgenic/pharmacological reduction of *HSP90* genes/HS proteins stimulate an increase of PA and acetylated-PA levels (2), a positive regulation of *AtPAOs* (3), and further increase of the PAO-related H_2O_2 (4). Reciprocally, underexpression of *AtPAO* in *Atpao3* mutant reveals stimulation of *AtHSP90* transcripts and HSP90 protein levels (5).

Taken all together, these results reveal a mutually antagonistic HS-regulated effect involving HSP90s, PAs, PAOs and H_2O_2 in *Arabidopsis thaliana*.

4. Materials and Methods

4.1. Plant Material and Treatments

Arabidopsis thaliana (ecotype Col-0, WT), an *pLFY:HSP90RNAi* silencing line against the four cytosolic *AtHSP90* genes [22], two T-DNA insertion mutant lines for the *Athsp90-1* and *Athsp90-4* (SALK_007614 and SALK_084059, respectively) and a T-DNA loss-of-function *Atpao3* (SALK_121288) line were used. In vitro produced seedlings were germinated on $\frac{1}{2}$ MS medium [84] supplemented with 0.5% sucrose on Petri dishes and plants were cultured in pots in a controlled growth chamber (23 °C and 8/16 h photoperiod under 150 µmol m^{-2} s^{-1}). Fully expanded leaves were excised from 4-week-old vegetative plants, before the floral transition phase, incubated in MES medium (pH 5.7) with agitation for 24 h, and subjected to different treatments: controls were incubated at 23 °C and HS samples at 42 °C, in the presence or absence of 10 mM Spd, 50 µM guazatine (specific inhibitor of PAO), 300 nM radicicol (specific inhibitor of HSP90) or 10 mM H_2O_2. Each chemical was added to the medium accompanied by brief vacuum treatment, 10 min prior to HS. All experiments were performed twice in identical growth chambers and growth conditions, with 4 replicates per treatment randomly designed. Samples were collected after 1 h treatment. Eight replicates from two independent experiments were pooled for statistical analysis. Asterisks indicate statistically significant differences from the WT control ($p < 0.05$).

4.2. Construction of HSP90-RNAi Line (pLFY:HSP90RNAi Mutant)

The construction of the *HSP90-RNAi* line is described in [22]. In this work, we analyzed the *pLFY:HSP90RNAi* mutant, line 10.1

4.3. RNA Extraction, RT-PCR and Semi-Q RT-PCR Analysis

Total RNA was extracted using the TRIZOL method (modified from Chomczinsky and Sacchi, [85]), any DNA contamination was removed with RQ1 RNase free DNase (Promega, Madison, WI, USA). About 1 µg was used as template in first-strand cDNA synthesis using PrimeScriptTM 1st Strand cDNA Synthesis Kit (TaKaRa, Kusatsu, Shiga, Japan), according to the manufacturer's instructions. RT-PCR Transcript analysis was performed using the gene specific primers, sequences relative sizes for RT-PCR are: *AtPAO1*: 468 bp, *AtPAO3*: 624 bp, *AtPAO5*: 269 bp, *AtHSP90-1*: 725 bp, *AtHSP90-2*: 197 bp, *AtHSP90-3*: 352 bp and *AtHSP90-4*: 234 bp (Table S1).

For the semi-Q RT-PCR, 5, 10 and 15 day-old seedlings of *Arabidopsis* wild type (Col) and *Atpao3* were used. First strand cDNA was synthesized using total RNA from each plant sample using SuperScript II Reverse Transcriptase (Invitrogen) in a 20 µL reaction. PCR amplification for each transcript (*AtHSP90-1*: 392 bp, *AtHSP90-2*: 192 bp, *AtHSP90-3*: 2531 bp and the *AtHSP90-4*: 128 bp) was performed using the gene-specific primers (Table S1). RT-PCR and semi-Q RT-PCR reactions were normalized using the *Ubq* and *18S* rDNA *Arabidopsis* genes, respectively [86]. For all semi-Q RT-PCRs, three technical replicates were prepared from materials pooled from all biological replicates/experiments. At least three variations in cycle number were used to verify reproducibility and amplification in the logarithmic phase, respectively. To estimate the amount of PCR product obtained, PCR reactions were run on ethidium bromide-containing agarose gels and subsequently analyzed in UV light. During gel analysis and data collection, volumes and settings were chosen to avoid saturation of fluorescence signals. RT and semi-Q RT-PCR products were quantified by measuring the intensity per unit area with Image J. The gene specificity of RT-PCR products was confirmed by sequencing. Data represent the mean of the technical replicates.

4.4. Determination and In Situ Localization of Hydrogen Peroxide

Hydrogen peroxide was quantified *via* a chemiluminescence assay [87]. In situ H_2O_2 was detected by DAB staining using the method of Thordal et al. [88].

4.5. Extraction and Quantification of PAs

Polyamines were extracted and analyzed according to Kotzabasis et al. [89], using an HP 1100 high performance liquid chromatographer (HPLC; Hewlett-Packard, Wadbronn, Germany).

4.6. Protein Extraction and Enzyme Assays

Total proteins were extracted as described in Papadakis and Roubelakis-Angelakis [90]. Protein content was measured according to the Lowry method [91]. PAO activity was measured according to the method of Yoda et al. [43] with minor modifications; fifty µg protein extract were incubated with 10 mM Spd for 30 min and measurement of generated H_2O_2 was performed for 1 min using a luminometer in a Tris-HCl (pH 8.0) reaction buffer containing 125 mM luminol [92].

4.7. Western Blotting

Protein extracts were prepared with Laemmli buffer and 50 µg of denaturated protein aliquots were loaded on 10% (*v/v*) polyacrylamide gel and electroblotted onto nitrocellulose membrane. Target proteins were then immunodetected against specific antibodies: a rabbit polyclonal anti-AtPAO3 anti-serum at a 1:10,000 dilution [39] and a monoclonal mouse anti-HSP90 at a 1:2000 dilution (Anti-HSP90 Mouse mAb, AC88, Calbiochem). For anti-AtPAO3 and anti-AtHSP90 horseradish peroxidase-conjugated anti-rabbit IgG (1:5,000) (Sigma, St Louis, MI, USA), or anti mouse IgG (1:10,000) (Sigma, St Louis, MI, USA) were used respectively. Western blots were developed using the Amersham ECL Prime Western Blotting Detection Reagent Kit (Pittsburg, CA, USA).

4.8. Image and Statistical Analysis

Image analysis was performed using ImageJ v 1.41 software, and statistical analysis using one-way ANOVA and the Duncan's test in the STATISTICA software.

Supplementary Materials: The following are available online at http://www.mdpi.com/2223-7747/8/9/323/s1, Figure S1: Soluble hydrolyzed (SH) PA fractions in leaves of 35 day-old WT and *athsp90-1*, *athsp90-4* and *pLFY;HSP90RNAi* plants, Figure S2: Western blot, Figure S3: Inhibition of PAO leaves cultured in MES, treated with HS, in the presence/absence of guazatine (Guaz) as a specific inhibitor induces, Figure S4: Abudance of mRNA of *AtHSP90-1*, *AtHSP90-2*, *AtHSP90-3*, *AtHSP90-4* genes in 5, 10 and 15 day-old WT and *Atpao3* mutant seedings, Figure S5: Abundance of mRNA, Table S1: List of specific primers for RT-PCR.

Author Contributions: Conceptualization, I.T. and K.A.R.-A.; data curation, I.T., M.G.P., T.M. and D.M.; formal analysis, I.T. and K.A.R.-A.; funding acquisition, K.A.R.-A.; investigation, I.T. and K.A.R.-A.; methodology, I.T., M.G.P. and K.A.R.-A.; project administration, K.A.R.-A.; resources, I.T., T.M. and K.A.R.-A.; supervision, D.M. and K.A.R.-A.; validation, K.A.R.-A.; writing—original draft, I.T.; writing—review & editing, D.M. and K.A.R.-A.

Funding: This work was funded by the EU and the Greek National funds, Research Funding Program THALES (MIS 377281 to K.A.R.-A.) and implemented in the frame of the EU COST FA1106 Action.

Conflicts of Interest: The authors declare no conflict of interest.

References

1. Song, H.; Zhao, R.; Fan, P.; Wang, X.; Chen, X.; Li, Y. Overexpression of *AtHsp90.2*, *AtHsp90.5* and *AtHsp90.7* in *Arabidopsis thaliana* enhances plant sensitivity to salt and drought stresses. *Planta* **2009**, *229*, 955–964. [CrossRef] [PubMed]
2. Lindquist, S.; Craig, E.A. The heat-shock proteins. *Annu. Rev. Inc.* **1988**, *22*, 631–677. [CrossRef] [PubMed]
3. De Maio, A.; Santoro, M.G.; Tanguay, R.M.; Hightower, L.E. Ferruccio Ritossa's scientific legacy 50 years after his discovery of the heat shock response: A new view of biology, a new society, and a new journal. *Cell Stress Chaperones.* **2012**, *17*, 139–143. [CrossRef] [PubMed]

4.   Zhang, J.; Srivastava, V.; Stewart, J.M.D. Molecular biology and physiology: Heat-tolerance in cotton is correlated with induced overexpression of heat-shock factors, heat-shock proteins, and general stress response genes. *J. Cotton Sci.* **2016**, *20*, 253–262.

5.   Queitsch, C.; Sangster, T.A.; Lindquist, S. Hsp90 as a capacitor of phenotypic variation. *Nature* **2002**, *417*, 618–624. [CrossRef] [PubMed]

6.   Song, H.; Fan, P.; Shi, W.; Zhao, R.; Li, Y. Expression of five AtHsp90 genes in *Saccharomyces cerevisiae* reveals functional differences of AtHsp90s under abiotic stresses. *J. Plant Physiol.* **2010**, *167*, 1172–1178. [CrossRef]

7.   Hahn, A.; Bublak, D.; Schleiff, E.; Scharf, K.D. Crosstalk between Hsp90 and Hsp70 chaperones and stress treanscription factors in tomato. *Plant Cell* **2011**, *23*, 741–755. [CrossRef] [PubMed]

8.   Bernfur, K.; Rutsdottir, G.; Emanuelsson, C. The chloroplast-localized small heat shock protein Hsp21 associates with the thylakoid membranes in heat-stressed plants. *Protein Sci.* **2017**, *26*, 1773–1784. [CrossRef]

9.   Kriechbaumer, V.; von Löffelholz, O.; Abell, B.M. Chaperone receptors: Guiding proteins to intracellular compartments. *Protoplasma* **2012**, *249*, 21–30. [CrossRef]

10.   Wang, R.; Zhang, Y.; Kieffet, M.; Yu, H.; Kepinski, S.; Estelle, M. HSP90 regulates temperature-dependent seedling growth by stabilizing the auxin co-receptor F-box protein TIR1. *Nat. Commun.* **2016**, *7*, 10269. [CrossRef]

11.   Crookes, W.J.; Olsen, L.J. The effects of chaperones and the influence of protein assembly on peroxisomal protein import. *J. Biol. Chem.* **1998**, *273*, 17236–17242. [CrossRef] [PubMed]

12.   Milioni, D.; Hatzopoulos, P. Genomic organization of *HSP90* gene family in *Arabidopsis*. *Plant Mol. Biol.* **1997**, *35*, 955–961. [CrossRef] [PubMed]

13.   Haralampidis, K.; Milioni, D.; Rigas, S.; Hatzopoulos, P. Combinatorial interaction of cis elements specifies the expression of the *Arabidopsis AtHsp90-1* gene. *Plant Physiol.* **2002**, *129*, 1138–1149. [CrossRef] [PubMed]

14.   Prasinos, C.; Krampis, K.; Samakovli, D.; Hatzopoulos, P. Tight regulation of expression of two Arabidopsis cytosolic *HSP90* genes during embryo development. *J. Exp. Bot.* **2004**, *56*, 633–644. [CrossRef] [PubMed]

15.   Sangster, T.A.; Bahrami, A.; Wilczek, A.; Watanabe, E.; Schellenberg, K. Phenotypic diversity and altered environmental plasticity in *Arabidopsis thaliana* with reduced Hsp90 levels. *PLoS ONE* **2007**, *2*, e684. [CrossRef]

16.   Prassinos, C.; Haralampidis, K.; Milioni, D.; Samakovli, D.; Krambis, K.; Hatzopoulos, P. Complexity of Hsp90 in organelle targeting. *Plant Mol. Biol.* **2008**, *67*, 323–334. [CrossRef]

17.   Krishna, P.; Gloor, G. The Hsp90 family of proteins in *Arabidopsis thaliana*. *Cell Stress Chaperones* **2001**, *6*, 238–246. [CrossRef]

18.   Yamada, K.; Nishimura, M. Cytosolic heat shock protein 90 regulates heat shock transcription factor in *Arabidopsis thaliana*. *Plant Signal. Behav.* **2008**, *3*, 660–662. [CrossRef]

19.   Xu, Z.S.; Li, Z.Y.; Chen, Y.; Chen, M.; Li, L.C.; Ma, Y.Z. Heat shock protein 90 in plants: Molecular mechanisms and roles in stress responses. *Int. J. Mol. Sci.* **2012**, *13*, 15706–15723. [CrossRef]

20.   Samakovli, S.; Thanou, A.; Valmas, C.; Hatzopoulos, P. Hsp90 canalizes developmental perturbation. *J. Exp. Bot.* **2007**, *58*, 3513–3524. [CrossRef]

21.   Kotak, S.; Vierling, E.; Baumlein, H.; von Koskull-Doring, P. A novel transcriptional cascade regulating expression of heat stress proteins during seed development of *Arabidopsis*. *Plant Cell* **2007**, *19*, 182–195. [CrossRef] [PubMed]

22.   Margaritopoulou, T.; Kryovrysanaki, N.; Megkoula, P.; Prasinos, C.; Samakovli, D.; Milioni, D.; Hatzopoulos, P. HSP90 canonical content organizes a molecular scaffold mechanism to progress flowering. *Plant J.* **2016**, *87*, 174–187. [CrossRef] [PubMed]

23.   Bao, F.; Huang, X.; Zhu, C.; Zhang, X.; Li, X.; Yang, S. Arabidopsis HSP90 protein modulates RPP4-mediated temperature-dependent cell death and defense responses. *New Phytol.* **2014**, *202*, 1320–1334. [CrossRef] [PubMed]

24.   Shigeta, T.; Zaizen, Y.; Asami, T.; Yoshida, S.; Nakamura, Y.; Okamoto, S.; Matsuo, T.; Sugimoto, Y. Molecular evidence of the involvement of heat shock protein 90 in brassinosteroid signaling in Arabidopsis T87 cultured cells. *Plant Cell Rep.* **2014**, *33*, 499–510. [CrossRef] [PubMed]

25.   Samakovli, D.; Margaritopoulou, T.; Prassinos, C.; Milioni, D.; Hatzopoulos, P. Brassinosteroid nuclear signaling recruits HSP90 activity. *New Phytol.* **2014**, *203*, 743–757. [CrossRef] [PubMed]

26.   Samakovli, D.; Ticha, T.; Ovecka, M.; Luptovciak, I.; Zapletalova, V.; Krasylenko, Y.; Komis, G.; Samajova, O.; Margaritopoulou, T.; Roka, L.; et al. Environment and HSP90 modulate MAPK stomatal developmental pathway. *BioRxiv* **2018**, 426684. [CrossRef]

27. Inoue, H.; Li, M.; Schnell, D.J. An essential role for chloroplast heat shock protein 90 (Hsp90C) in protein import into chloroplasts. *PNAS* **2013**, *110*, 3173–3178. [CrossRef] [PubMed]

28. Kindgren, P.; Noren, L.; de Dios, J.; Lopez, B.; Shaikhali, J.; Strand, A. Interplay between Heat Shock Protein 90 and HY5 controls PhANG expression in response to the GUN5 plastid signal. *Mol. Plant* **2012**, *5*, 901–913. [CrossRef] [PubMed]

29. Kadota, Y.; Shirasu, K.; Guerios, R. NLR sensors meet at the SGT1-HSP90 crossroad. *Trends Biochem. Sci.* **2010**, *35*, 199–207. [CrossRef]

30. Miernyk, J.A. Protein folding in the Plant Cell. *Plant Physiol.* **1999**, *121*, 695–703. [CrossRef]

31. Hong, S.W.; Vierling, E. Mutants of Arabidopsis thaliana defective in the acquisition of tolerance to high temperature stress. *PNAS* **2000**, *97*, 4392–4397. [CrossRef] [PubMed]

32. Wang, W.; Vinocur, B.; Shoseyov, O.; Altman, A. Role of plant heat-shock proteins and molecular chaperones in the abiotic stress response. *Trends Plant Sci.* **2004**, *9*, 244–252. [CrossRef] [PubMed]

33. Kozeko, L.Y. The role of HSP90 chaperones in stability and plasticity of ontogenesis of plants under normal and stressful conditions (Arabidopsis thaliana). *Cytol. Genet.* **2019**, *2019. 53*, 143–161. [CrossRef]

34. Clark, C.B.; Rane, M.J.; El Mehdi, D.; Miller, C.J.; Sachleben, L.R., Jr.; Gozal, E. Role of oxidative stress in geldanamycin-induced cytotoxicity and disruption of Hsp90 signaling complex. *Free Radic. Biol. Med.* **2009**, *47*, 1440–1449. [CrossRef] [PubMed]

35. Tiburcio, A.F.; Altabella, T.; Bitrián, M.; Alcázar, R. The roles of Polyamines during the lifespan of plants: from development to stress. *Planta* **2014**, *240*, 1–18. [CrossRef] [PubMed]

36. Bagni, N.; Tassoni, A. Biosynthesis, oxidation and conjugation of aliphatic Polyamines in higher plants. *J Amino Acids* **2001**, *20*, 301–317. [CrossRef]

37. Moschou, P.N.; Delis, I.D.; Paschalidis, K.A.; Roubelakis-Angelakis, K.A. Transgenic tobacco plants overexpressing poliamine oxidase are not able to cope with oxidative burst generated by abiotic factors. *Physiol. Plant* **2008**, *133*, 140–156. [CrossRef]

38. Moschou, P.N.; Paschalidis, K.A.; Delis, I.D.; Andriopoulou, A.H.; Lagiotis, G.D.; Yakoumakis, D.I.; Roubelakis-Angelakis, K.A. Spermidine exodus and oxidation in the apoplast induced by abiotic stress is responsible for H_2O_2 signatures that direct tolerance responses in tobacco. *Plant Cell* **2008**, *20*, 1708–1724. [CrossRef]

39. Moschou, P.N.; Sanmartin, M.; Andriopoulou, A.H.; Rojo, E.; Sanchez-Serrano, J.J.; Roubelakis-Angelakis, K.A. Bridging the gap between plant and mammalian polyamine catabolism: A novel peroxisomal polyamine oxidase responsible for a full back-conversion pathway in *Arabidopsis*. *Plant Physiol.* **2008**, *147*, 1845–1857. [CrossRef]

40. Paschalidis, K.A.; Roubelakis-Angelakis, K.A. Spatial and temporal distribution of polyamine levels and polyamine anabolism in different organs/tissues of the tobacco plant: Correlations with Age, Cell Division/Expansion, and Differentiation. *Plant Physiol.* **2005**, *138*, 142–152. [CrossRef]

41. Paschalidis, K.A.; Roubelakis-Angelakis, K.A. Sites and regulation of polyamine catabolism in the tobacco plant. Correlations with cell division/expansion, cell-cycle progression, and vascular development. *Plant Physiol.* **2005**, *138*, 2174–2184. [CrossRef] [PubMed]

42. Mattoo, A.K.; Minocha, S.C.; Minocha, R.; Handa, A.K. Polyamines and cellular metabolism in plants: Transgenic approaches reveal different responses to diamine putrescine versus higher Polyamines spermidine and spermine. *Amino Acids* **2010**, *38*, 405–413. [CrossRef] [PubMed]

43. Yoda, H.; Yamaguchi, Y.; Sano, H. Induction of hypersensitive cell death by Hydrogen Peroxide produced through polyamine degradation in tobacco plants. *Plant Physiol.* **2003**, *132*, 1973–1981. [CrossRef] [PubMed]

44. Yamaguchi, K.; Takahashi, Y.; Berberich, T.; Imai, A.; Takahashi, T.; Michael, A.J.; Kusano, T. A Protective role for the polyamine spermine against drought stress in Arabidopsis. *Biochem. Biophys. Res. Commun.* **2007**, *352*, 486–490. [CrossRef] [PubMed]

45. Fincato, P.; Moschou, P.N.; Spedaletti, V.; Tavazza, R.; Angelini, R.; Federico, R.; Roubelakis-Angelakis, K.A.; Tavladoraki, P. Functional diversity inside the Arabidopsis polyamine oxidase gene family. *J. Exp. Bot.* **2011**, *62*, 1155–1168. [CrossRef] [PubMed]

46. Fincato, P.; Moschou, P.N.; Ahou, A.; Angelini, R.; Roubelakis-Angelakis, K.A.; Federico, R.; Tavladoraki, P. The members of *Arabidopsis thaliana* PAO gene family exhibit distinct tissue- and organ-specific expression pattern during seedling growth and flower development. *Amino Acids* **2012**, *42*, 831–841. [CrossRef] [PubMed]

47. Tisi, A.; Federico, R.; Moreno, S.; Lucretti, S.; Moschou, P.N.; Roubelakis-Angelakis, K.A.; Angelini, R.; Cona, A. Perturbation of polyamine catabolism can strongly affect root development and xylem differentiation. *Plant Physiol.* **2011**, *157*, 200–215. [CrossRef]

48. Kudou, M.; Shiraki, K.; Fujiwara, S.; Imanaka, T.; Takagi, M. Prevention of thermal inactivation and aggregation of lysozymes by Polyamines. *Eur. J. Biochem.* **2003**, *270*, 4547–4554. [CrossRef]

49. Königshofer, H.; Lechner, S. Are Polyamines involved in the synthesis of heat-shock proteins in cell suspension cultures of tobacco and alfalfa in response to high-temperatture stress? *Plant Physiol. Biochem.* **2002**, *40*, 51–59. [CrossRef]

50. Moschou, P.N.; Roubelakis-Angelakis, K.A. Polyamines and programmed cell death. *J. Exp. Bot.* **2013**, *65*, 1285–1296. [CrossRef]

51. Tavladoraki, P.; Cona, A.; Federico, R.; Tempera, G.; Viceconte, N.; Saccocio, S.; Battaglia, V.; Toninello, A.; Agostinelli, E. Polyamine catabolism: Target for antiproliferative therapies in animals and stress tolerance strategies in plants. *Amino Acids* **2012**, *42*, 411–426. [CrossRef] [PubMed]

52. Yoda, H.; Fujimura, K.; Takahashi, H.; Munemura, I.; Uchimiya, H.; Sano, H. Polyamines as a common source of hydrogen peroxide in host- and nonhost hypersensitive response during pathogen infection. *Plant Mol. Biol.* **2009**, *70*, 103–112. [CrossRef] [PubMed]

53. Tavladoraki, P.; Rossi, M.N.; Saccuti, G.; Perez-Amador, M.A.; Polticelli, F.; Angelini, R.; Federico, R. Heterologous expression and biochemical characterization of a polyamine oxidase from Arabidopsis involved in polyamine back conversion. *Plant Physiol.* **2006**, *141*, 1519–1532. [CrossRef] [PubMed]

54. Kamada-Nobusada, T.; Makoto, H.; Fukazawa, M.; Sakakibara, H.; Nishimura, M. A putative peroxisomal polyamine oxidase, *AtPAO4*, is involved in polyamine catabolism in *Arabidopsis thaliana*. *Plant Cell Physiol.* **2008**, *49*, 1272–1282. [CrossRef] [PubMed]

55. Angelini, R.; Cona, A.; Federico, R.; Fincato, P.; Tavladoraki, P.; Tisi, A. Plant amine oxidases "on the move": An update. *Plant Physiol. Biochem.* **2010**, *48*, 560–564. [CrossRef] [PubMed]

56. Moschou, P.N.; Roubelakis-Angelakis, K.A. Characterization, assay, and substrate specificity of plant polyamine oxidases. *Meth. Mol. Biol.* **2001**, *720*, 183–194.

57. Pantano, C.; Shrivastava, P.; McElhinney, B.; Janssen-Heininger, Y. Hydrogen peroxide signaling through tumor necrosis factors receptor 1 leads to selective activation of c-Jun N-terminal kinase. *J. Biol. Chem.* **2003**, *278*, 44091–44096. [CrossRef]

58. Baxter, A.; Mittler, R.; Suzuki, N. ROS as key players in plant stress signalling. *J. Exp. Bot.* **2004**, *65*, 1229–1240. [CrossRef]

59. Ghuge, S.A.; Tisi, A.; Carucci, A.; Rodrigues-Pousada, R.A.; Franchi, S.; Tavladoraki, P.; Angelini, R.; Cona, A. Cell wall amine oxidases: New players in root xylem differentiation under stress conditions. *Plants* **2015**, *4*, 489–504. [CrossRef]

60. Wang, W.; Paschalidis, K.; Feng, J.C.; Song, J.; Liu, J.H. Polyamine catabolism in plants: A universal process with diverse functions. *Front. Plant Sci.* **2019**, *10*. [CrossRef]

61. Gémes, K.; Kim, Y.J.; Park, K.Y.; Moschou, P.N.; Andronis, E.; Valassakis, C.; Roussis, A.; Roubelakis-Angelakis, K.A. A NADPH-Oxidase/polyamine oxidase feedback loop controls oxidative burst under salinity. *Plant Physiol.* **2016**, *172*, 1418–1431. [CrossRef] [PubMed]

62. Casero, R.A., Jr.; Pegg, A.E. Spermidine/spermine N1-acetyltransferase—The turning point in polyamine metabolism. *FASEB J.* **1991**, *7*, 653–661.

63. Bardocz, S.; White, A. Effect of lectins on uptake of Polyamines. *Meth. Mol. Med.* **1998**, *9*, 393–405.

64. Desiderio, M.A.; Dansi, P.; Tacchini, L.; Bernelli-Zazzera, A. Influence of Polyamines on DNA binding of heat shock and activator protein 1 transcription factors induced by heat shock. *FEBS Lett.* **1999**, *455*, 149–153. [CrossRef]

65. Sagor, G.H.; Berberich, T.; Takahashi, Y.; Niitsu, M.; Kusano, T. The polyamine spermine protects *Arabidopsis* from heat stress-induced damage by increasing expression of heat shock-related genes. *Transgenic Res.* **2013**, *22*, 595–605. [CrossRef] [PubMed]

66. Mellidou, I.; Karamanoli, K.; Berris, D.; Haralampidis, K.; Constantinidou, H.I.; Roubelakis-Angelakis, K.A. Underexpression of apoplastic polyamine oxidase improves thermotolerance in *Nicotiana tabacum*. *J. Plant Physiol.* **2017**, *218*, 171–174. [CrossRef] [PubMed]

67. Adio, A.M.; Casteel, C.L.; de Vos, M.; Kim, J.H.; Joshi, V.; Li, B.; Juéry, C.; Daron, J.; Kliebenstein, D.J.; Jander, G. Biosynthesis and defensive function of Nδ-acetylornithine, a jasmonate-induced Arabidopsis metabolite. *Plant Cell* **2011**, *23*, 3303–3318. [CrossRef] [PubMed]

68. Lou, Y.R.; Bor, M.; Yan, J.; Preuss, A.S.; Jander, G. Arabidopsis NATA1 acetylates Putrescine and decreases defense-related hydrogen peroxide accumulation. *Plant Physiol.* **2016**, *171*, 1443–1455. [CrossRef] [PubMed]

69. Tassoni, A.; van Buuren, M.; Franceschetti, M.; Fornalè, S.; Bagni, N. Polyamine content and metabolism in Arabidopsis thaliana and effect of spermidine on plant development. *Plant Physiol. Biochem.* **2000**, *38*, 383–393. [CrossRef]

70. Kee, K.; Foster, B.A.; Merali, S.; Kramer, D.L.; Hensen, M.L.; Diegelman, P.; Kisiel, N.; Vujcic, S.; Mazurchuk, R.V.; Porter, C.W. Activated polyamine catabolism depletes acetyl-CoA pools and suppresses prostate tumor growth in TRAMP mice. *J. Biol. Chem.* **2004**, *279*, 40076–40083. [CrossRef] [PubMed]

71. Pegg, A.E. Spermidine/Spermine-N^1-acetyltransferase: A key metabolic regulator. *Am. J. Physiol. Endocrinol. Metab.* **2008**, *294*, 995–1010. [CrossRef] [PubMed]

72. Tian, Y.; Wang, S.; Wang, B.; Zhang, J.; Jiang, R.; Zhang, W. Overexpression of *SSAT* by DENSPM treatment induces cell detachment and apoptosis in glioblastoma. *Oncol. Rep.* **2012**, *27*, 1227–1232. [CrossRef] [PubMed]

73. Fliniaux, O.; Mesnard, F.; Raynaud-Le Grandic, S.; Baltora-Rosset, S.; Bienaimé, C.; Robins, R.J.; Fliniaux, M.A. Altered nitrogen metabolism associated with de-differentiated suspension cultures derived from root cultures of Datura stramonium studied by heteronuclear multiple bond coherence (HMBC) NMR spectroscopy. *J. Exp. Bot.* **2004**, *55*, 1053–1060. [CrossRef] [PubMed]

74. Alcázar, R.; Planas, J.; Saxena, T.; Zarza, X.; Bortolotti, C.; Cuevas, J.; Bitrián, M.; Tiburcio, A.F.; Altabella, T. Putrescine accumulation confers drought tolerance in transgenic Arabidopsis plants over-expressing the homologous Arginine decarboxylase 2 gene. *Plant Physiol. Biochem.* **2010**, *48*, 547–552. [CrossRef] [PubMed]

75. Chen, F.; Pandey, D.; Chadi, A.; Catravas, J.D.; Chen, T.; Fulton, D.J. Hsp90 regulates NADPH oxidase activity and its necessary for superoxide but not hydrogen peroxide production. *Antioxid. Redox. Signal* **2011**, *14*, 2107–2119. [CrossRef] [PubMed]

76. Xu, X.; Song, H.; Zhou, Z.; Shi, N.; Ying, Q.; Wang, H. Functional characterization of AtHsp90.3 in *Saccharomyces cerevisiae* and *Arabidopsis thaliana* under heat stress. *Biotechnol Lett.* **2010**, *32*, 979–987. [CrossRef] [PubMed]

77. Wiech, H.; Buchner, J.; Zimmermann, R.; Jakob, U. Hsp90 chaperones protein folding in vitro. *Nature* **1992**, *358*, 169–170. [CrossRef]

78. Wang, C.; Gu, X.; Wang, X.; Guo, H.; Geng, J.; Yu, H.; Sun, J. Stress response and potential biomarkers in spinach (*Spinacia oleracea* L.) seedlings exposed to soil lead. *Ecotoxicol. Environ. Saf.* **2011**, *74*, 41–47. [CrossRef]

79. Nishizawa-Yokoi, A.; Tainaka, H.; Yoshida, E.; Tamoi, M.; Yabuta, Y.; Shigeoka, S. The 26S Proteasome Function and Hsp90 activity involved in the regulation of HsfA2 expression in response to oxidative stress. *Plant Cell Physiol.* **2010**, *51*, 486–496. [CrossRef]

80. McLellan, C.A.; Turbyvilee, T.J.; Wijeratne, E.K.; Kerschen, A.; Vierling, E.; Queitsch, C.; Whitesell, L.; Gunatilaka, A.L. A Rhizosphere fungus enhances Arabidopsis thermotolerance through production of an HSP90 inhibitor. *Plant Physiol.* **2007**, *145*, 174–182. [CrossRef]

81. Kim, D.W.; Watanabe, K.; Murayama, C.; Izawa, S.; Niitsu, M.; Michael, A.J.; Berberich, T.; Kusano, T. Polyamine oxidase 5 regulates Arabidopsis growth through Thermospermine Oxidase activity. *Plant Physiol.* **2014**, *165*, 1575–1590. [CrossRef] [PubMed]

82. Paschalidis, A.K.; Toumi, I.; Moschou, N.P.; Roubelakis-Angelakis, K.A. ABA-dependent amine oxidases-derived H$_2$O$_2$ affects stomata conductance. *Plant Signal. Behav.* **2010**, *5*, 1153–1156.

83. Andronis, E.A.; Moschou, P.N.; Toumi, I.; Roubelakis-Angelakis, K.A. Peroxisomal polyamine oxidase and NADPH-oxidase cross-talk for ROS homeostasis which affects respiration rate in *Arabidopsis thaliana*. *Front. Plant Sci.* **2014**, *3*, 132. [CrossRef] [PubMed]

84. Murashige, T.; Skoog, F. A revised medium for rapid growth and bioassays with tobacco tissue culture. *Physiol. Plant.* **1962**, *15*, 473–497. [CrossRef]

85. Chomczinsky, P.; Sacchi, N. Single-step method of RNA isolation by acid guanidinium thiocyanate-phenol-chloroform extraction. *Anal. Biochem.* **1987**, *162*, 156–159. [CrossRef]

86. Galeano, E.; Vasconcelos, T.S.; Ramiro, D.A.; de Martin, V.d.F.; Carrer, H. Identification and validation of quantitative real-time reverse transcription PCR reference genes for gene expression analysis in Teak (Tectona grandis L.f.). *BMC Res. Notes* **2014**, *7*, 464. [CrossRef] [PubMed]

87. Papadakis, A.K.; Roubelakis-Angelakis, K.A. The regeneration of active oxygen species differs in tobacco and grapevine mesophyll protoplasts. *Plant Physiol.* **1999**, *121*, 197–206. [CrossRef]

88. Thordal-Christensen, H.; Zhang, Z.; Wei, Y.; Collinge, D.B. Subcellular localization of H_2O_2 in plants. H_2O_2 accumulation in papillae and hypersensitive response during the barley-powdery mildew interaction. *Plant J.* **1997**, *11*, 1187–1194. [CrossRef]

89. Kotzabasis, K.; Christakis-Hampsas, M.D.; Roubelakis-Angelakis, K.A. A narrow-bore HPLC method for the identification and quantitation of free, conjugated, and bound Polyamines. *Analyt. Biochem.* **1993**, *214*, 484–489. [CrossRef]

90. Papadakis, A.K.; Roubelakis-Angelakis, K.A. Polyamines inhibit NADPH oxidase-mediated superoxide generation and putrescine prevents programmed cell death induced by polyamine oxidase-generated hydrogen peroxide. *Planta* **2005**, *220*, 826–837. [CrossRef]

91. Lowry, O.H.; Rosebrough, N.J.; Farr, A.L.; Randall, R.J. Protein measurement with the Folin phenol reagent. *J. Biol. Chem.* **1951**, *193*, 265–275. [PubMed]

92. Bolwell, G.P.; Butt, V.S.; Davies, D.R.; Zimmerlin, A. The origin of the oxidative burst in plant cells. *Free Radic. Res.* **1995**, *23*, 517–532. [CrossRef] [PubMed]

Article

The Copper Amine Oxidase AtCuAOδ Participates in Abscisic Acid-Induced Stomatal Closure in Arabidopsis

Ilaria Fraudentali [1], Sandip A. Ghuge [2], Andrea Carucci [1], Paraskevi Tavladoraki [1,3], Riccardo Angelini [1,3], Alessandra Cona [1,3] and Renato A. Rodrigues-Pousada [4,*]

[1] Department of Science, Università Roma Tre, 00146 Roma, Italy; ilaria.fraudentali@uniroma3.it (I.F.); andrea.carucci@outlook.it (A.C.); paraskevi.tavladoraki@uniroma3.it (P.T.); riccardo.angelini@uniroma3.it (R.A.); alessandra.cona@uniroma3.it (A.C.)

[2] Institute of Plant Sciences, The Volcani Center, ARO, Bet Dagan 50250, Israel; sandip.ghuge.biotech@gmail.com

[3] Istituto Nazionale Biostrutture e Biosistemi (INBB), 00136 Rome, Italy

[4] Department of Life, Health, and Environmental Sciences, Università dell'Aquila, 67100 L'Aquila, Italy

* Correspondence: pousada@univaq.it; Tel.: +39-0862433268

Received: 10 May 2019; Accepted: 17 June 2019; Published: 20 June 2019

Abstract: Plant copper amine oxidases (CuAOs) are involved in wound healing, defense against pathogens, methyl-jasmonate-induced protoxylem differentiation, and abscisic acid (ABA)-induced stomatal closure. In the present study, we investigated the role of the *Arabidopsis thaliana* CuAOδ (AtCuAOδ; At4g12290) in the ABA-mediated stomatal closure by genetic and pharmacological approaches. Obtained data show that *AtCuAOδ* is up-regulated by ABA and that two *Atcuaoδ* T-DNA insertional mutants are less responsive to this hormone, showing reduced ABA-mediated stomatal closure and H_2O_2 accumulation in guard cells as compared to the wild-type (WT) plants. Furthermore, CuAO inhibitors, as well as the hydrogen peroxide (H_2O_2) scavenger N,N^1-dimethylthiourea, reversed most of the ABA-induced stomatal closure in WT plants. Consistently, *AtCuAOδ* over-expressing transgenic plants display a constitutively increased stomatal closure and increased H_2O_2 production compared to WT plants. Our data suggest that AtCuAOδ is involved in the H_2O_2 production related to ABA-induced stomatal closure.

Keywords: copper amine oxidases; H_2O_2; ROS; polyamines; ABA; stomatal closure

---

1. Introduction

Copper amine oxidases (CuAOs) are dimeric proteins of 140–180 kDa, containing a copper ion and a redox-active organic cofactor 2,4,5-trihydroxyphenylalanine quinone (TPQ) for each monomer. These enzymes catalyze the intracellular and extracellular terminal catabolism of amines, including monoamines, diamines, and polyamines (PAs), by oxidizing the carbon next to the primary amino group, with the subsequent reduction of molecular oxygen to hydrogen peroxide (H_2O_2) and the production of the corresponding aldehydes and ammonia [1,2]. CuAOs have been found at high expression levels in several species of *Fabaceae*, especially in the cell wall of pea (*Pisum sativum*), chickpea (*Cicer arietinum*), lentil (*Lens culinaris*), and soybean (*Glycine max*) seedlings, from which these enzymes have been purified and characterized [3]. CuAOs from these species preferentially oxidize the diamine putrescine (Put) and cadaverine [4]. A number of peroxisomal and apoplastic CuAOs have been described in Arabidopsis (*Arabidopsis thaliana*) [5–8], tobacco (*Nicotiana tabacum*) [8,9], and apple (*Malus domestica*) [10], and a latex CuAO has been characterized from Mediterranean spurge (*Euphorbia characias*) [11], with diverse substrate affinities and specificities [12].

CuAOs belong to the larger family of amine oxidases (AOs), which also includes flavin adenine dinucleotide (FAD)-dependent polyamine oxidases (PAOs) [3]. The latter enzymes oxidize PAs to aminoaldehydes at the carbon neighboring the secondary amino group, resulting in the production of different residual amine moieties, depending on the location of the oxidized carbon in the aliphatic chain [1,2,12]. The other major co-product of the PAO-catalyzed PA oxidation reaction is hydrogen peroxide (H_2O_2) [2].

In spite of differences in biochemical features, such as protein structure, co-factors, and catalytic mechanisms, CuAOs and PAOs partially share substrates and reaction products and play overlapping roles in both the intracellular control of PA's homeostasis and production of biologically active compounds and metabolites, such as the developmentally-controlled or stress-induced H_2O_2 [1–3]. Regarding this, the PA-derived apoplastic H_2O_2 has been proposed to act as both a signal for activation of defense gene expression and as a co-substrate for the peroxidase-driven reactions of wall-stiffening and lignification events [3]. In particular, plant AOs have been involved in a variety of growth and developmental events, including light-induced inhibition of mesocotyl growth [13], root xylem differentiation [14] and pollen tube growth [15], as well as in stress tolerance and defense responses, especially salt stress [16], wounding [17], pathogen attack [9,18–21], and stomatal closure [22–25].

Ten putative *CuAO* genes are annotated in the Arabidopsis genome, four of which have been characterized for substrate specificity and subcellular localization of the encoded enzymes and regulation of gene expression. The apoplastic AtCuAOβ (formerly AtAO1; At4g14940) [5] and AtCuAOγ1 (formerly AtCuAO1; At1g62810), the peroxisomal AtCuAOα3 (formerly AtCuAO2; At1g31710), and AtCuAOζ (formerly AtCuAO3; At2g42490) [7,12,25] all oxidize Spd at the primary amino group with an affinity comparable to that for Put. Expression of these AtCuAO-encoding genes is inducible by stress-related hormones and elicitors, such as methyl-jasmonate (MeJA; *AtCuAOβ, AtCuAOγ1, AtCuAOα3* and *AtCuAOζ*), abscisic acid (ABA), salicylic acid, and flagellin 22 (*AtCuAOγ1, AtCuAOζ*), and by wounding (*AtCuAOα3*) [7,26]. Notwithstanding the high number of annotated genes, only a few studies concerning the physiological roles of *AtCuAOs* have been reported so far. In this regard, it has been described that *AtCuAOγ1* and *AtCuAOζ* are involved in the ABA-mediated stress responses by contributing respectively to the ABA-induced production of nitric oxide (NO) [27] and the ABA-induced stomatal closure [25]. Furthermore, it has been shown that the AtCuAOβ-driven production of apoplastic H_2O_2 signals the MeJA-mediated protoxylem differentiation in Arabidopsis roots [26,28]. In this regard, *AtCuAOβ* gene expression in guard cells of leaves and flowers has been demonstrated, suggesting a role for this gene also in the control of stomatal closure [29]. Concerning the other *AtCuAO* annotated genes, the gene product of *AtCuAOδ* (At4g12290) has been identified among proteins purified from the central vacuoles of rosette leaf tissue by means of complementary proteomic methodologies [30].

The role played by the vacuole in ABA-induced stomatal closure [31], along with the occurrence of an ABA-inducible *AtCuAOδ* expression in guard cells, as reported by the Arabidopsis eFP Browser (http://bar.utoronto.ca/efp/cgi-bin/efpWeb.cgi; [32]), led us to analyze the possible involvement of the vacuolar AtCuAOδ in the control of stomatal movement. Herein we provide genetic and physiological evidence for a role of this protein as a H_2O_2 source in the ABA-induced stomatal closure.

2. Results

2.1. AtCuAOδ Expression Is Induced by ABA

A promoter region of approximately 2.7 kb upstream of the *AtCuAOδ* start codon was analyzed in silico for the presence of cis-acting elements by the Arabidopsis eFP Browser (http://bar.utoronto.ca/cistome/cgi-bin/BAR_Cistome.cgi). On the basis of this analysis, two recognition sequences (CATGTG) for the ABA-inducible MYC factor (MYCATERD1) necessary for the expression of *erd1* (early responsive to dehydration) in dehydrated Arabidopsis plants were identified. Moreover, the analysis of microarray data retrieved from the Arabidopsis eFP Browser revealed the occurrence of

AtCuAOδ mRNA in guard cells, whose level increased upon ABA-treatment. These data are supported by reverse transcription-quantitative polymerase chain reaction (RT-qPCR) studies that showed a two- to three-fold increase of *AtCuAOδ* expression levels depending on ABA concentration as soon as 3 h after the onset of treatment (Figure 1). This induction peaked at 6 h with a four-fold increase at 100 μM ABA, and returned to almost control levels at 24 h for the two lower concentrations while it was still two-fold higher at 100 μM ABA.

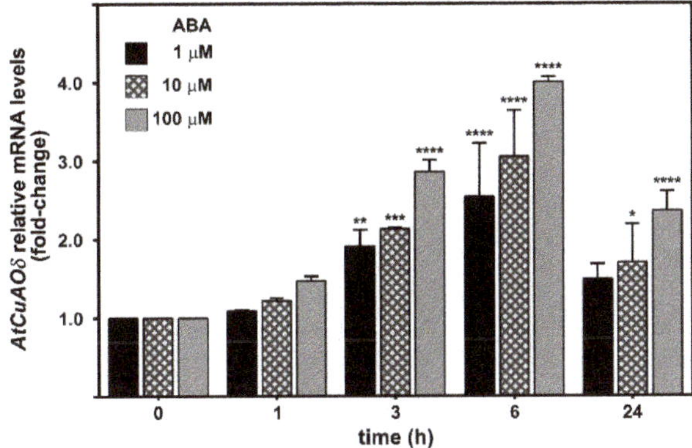

Figure 1. Analysis of *AtCuAOδ* gene expression upon abscisic acid (ABA) treatment by reverse transcription-quantitative polymerase chain reaction (RT-qPCR). The expression of *AtCuAOδ* gene was analyzed in 12-day-old wild-type (WT) seedlings untreated or treated with 1, 10, and 100 μM ABA for 0, 1, 3, 6, and 24 h. Five independent experiments as biological replicates (mean values ± SD; $n = 5$) were performed. *AtCuAOδ* mRNA level after ABA treatment is relative to that of the corresponding untreated plant for each time point. The significance levels between the relative mRNA level at each time and the mRNA level of control untreated plant at time 0, which is assumed to be one, is reported. *P* values have been calculated with one-way analysis of variance (ANOVA); *, **, ***, and **** *p* values equal or are less than 0.05, 0.01, 0.001, and 0.0001, respectively.

2.2. *AtCuAOδ* Loss-of-Function Mutants Are Unresponsive to ABA-Induced Stomatal Closure

In order to investigate the contribution of *AtCuAOδ* in ABA-mediated responses, two T-DNA insertional mutant lines for this gene [SALK 072954.55.00.x line, TAIR (The Arabidopsis Information Resource) accession number 4122972 and GK-011C04-013046 line, TAIR accession number 4242275] were identified from the TAIR database (http://www.arabidopsis.org/; [33]), and obtained, hereafter referred to as *Atcuaoδ.1* and *Atcuaoδ.2* (Figure S1). From evidence available in TAIR, the T-DNA insertion sites are located in the first exon in both mutants (Figure S1), upstream of the encoded catalytic site active residues (Figure S2), thereby creating loss-of-function mutants. Mutant plants homozygous for the T-DNA insertions were identified by PCR analysis of genomic DNA (Figure S1). RT-PCR analysis of the selected plants confirmed the absence of the full-length gene transcripts in *Atcuaoδ.1* and *Atcuaoδ.2* (Figure S1). Analysis of *Atcuaoδ* mutants under physiological growth conditions did not highlight any apparent different phenotypes, i.e., germination events, stem or root length, and leaf morphology (data not shown).

As ABA has well characterized effects in the regulation of the stomatal aperture, we investigated whether AtCuAOδ-driven PA oxidation in guard cells could be involved in the ABA-mediated stomatal closure. Wild-type (WT), *Atcuaoδ.1*, and *Atcuaoδ.2* plants were treated with ABA (1, 10, and 100 μM) for 2 h and the stomatal aperture was analyzed by measuring the width and the length of the stomatal pore (width/length ratio). In Figure 2A and Table S1, we show that while between control untreated WT (Control WT) and control untreated insertional mutants (Control *Atcuaoδ.1* or Control *Atcuaoδ.2*) no width/length ratio differences were detected, a significant reduction of the ABA-mediated stomatal closure was observed in the *Atcuaoδ* mutants as compared to WT plants. Indeed, stomatal closure was induced in WT by ABA treatments of about 51% at 1 μM and 77% at 100 μM as compared to Control WT, while the ABA-mediated stomatal closure in *Atcuaoδ.1* and in *Atcuaoδ.2* ranged from ~9–12% to a maximum of ~15–17%, respectively, as compared to the corresponding Control mutant plants (Figure 2A).

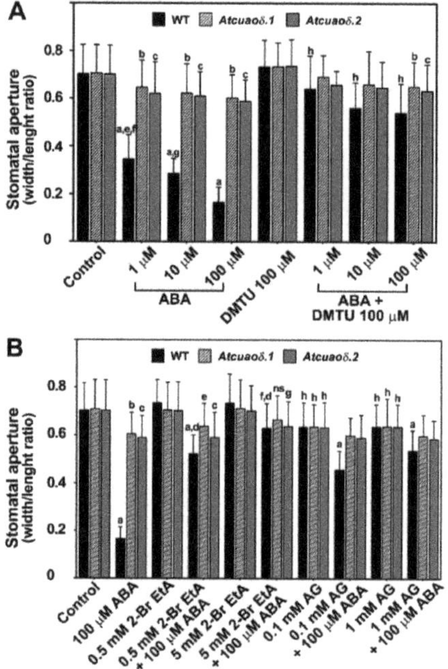

Figure 2. Effect of ABA, *N,N'*-dimethylthiourea (DMTU) (**A**) and CuAO inhibitors, 2-BrEtA and aminoguanidine (AG) (**B**), on stomatal pore width/length ratio of 12-day-old seedlings from WT, *Atcuaoδ.1*, and *Atcuaoδ.2*. Mean values ± SD (n = 15) are reported. *P* values have been calculated with one-way ANOVA analysis; non-significant (ns) differences: p values > 0.05; *, **, ***, and **** p values are equal to or less than 0.05, 0.01, 0.001, and 0.0001, respectively. (**A**) Seedlings were treated for 2 h with ABA (1, 10, and 100 μM) and DMTU (100 μM), either alone or in combination with the hormone. The significance levels are described with letters where appropriate; non-significant differences are not indicated; a = ****, ABA 1, 10, and 100 μM WT vs. Control WT; b = **, ABA 1 μM/ABA 100 μM + DMTU *Atcuaoδ.1* vs. Control *Atcuaoδ.1*; c = ****, ABA 1, 10, 100 μM *Atcuaoδ.2*/ABA 100 μM + DMTU *Atcuaoδ.2* vs. Control *Atcuaoδ.2*; d= ****, ABA 10, 100 μM *Atcuaoδ.1* vs. Control *Atcuaoδ.1*; e = *, ABA 1 μM WT vs. ABA 10 μM WT; f = ****, ABA 1 μM WT vs. ABA 100 μM WT; g = ****, ABA 10 μM WT vs. ABA 100 μM WT; h = ****, ABA 1, 10, and 100 μM WT vs. ABA 1, 10, and 100 μM + DMTU WT. (**B**) Seedlings were treated with 2-BrEtA (0.5, 5 mM) or AG (0.1, 1 mM) for 30 min. ABA was added (100 μM) and further incubated for 2 h. The significance levels are described with letters where appropriate.; ns: ABA 100 μM + 5 mM 2-BrEtA *Atcuaoδ.1* vs. Control *Atcuaoδ.1*; a: ****, ABA 100 μM, ABA 100 μM + 0.5 mM 2-BrEtA, ABA 100 μM + AG 0.1 or 1 mM WT vs. Control WT; b: ****, ABA 100 μM *Atcuaoδ.1* vs. Control *Atcuaoδ.1*; c: ****, ABA 100 μM/ABA 100 μM + 0.5mM 2-BrEtA *Atcuaoδ.2* vs. Control *Atcuaoδ.2*; d: ****, ABA 100 μM + 2-BrEtA 0.5 or 5 mM WT vs. ABA 100 μM WT; e: ** ABA 100 μM + 0.5 mM 2-BrEtA *Atcuaoδ.1* vs. Control *Atcuaoδ.1*; f: ** ABA 100 μM + 2-BrEtA 5 mM WT vs. Control WT; g: * ABA 100 μM + 2-BrEtA 5 mM *Atcuaoδ.2* vs. Control *Atcuaoδ.2*; h: *, AG 0.1 or 1 mM, WT, *Atcuaoδ.1* e *Atcuaoδ.2* vs. Control WT, Control *Atcuaoδ.1*, or Control *Atcuaoδ.2*.

Consistent with these data, treatment with the CuAO-specific inhibitors, 2-bromoethylamine (2-BrEtA) and aminoguanidine (AG), inhibited the ABA-mediated stomatal closure in WT plants (Figure 2B and Table S2). The action of the CuAO-specific inhibitors, 2-BrEtA and AG, on stomatal apertures of *Atcuaoδ* mutants was also studied. At the two different inhibitor concentrations used, in the presence of 100 μM ABA (the highest hormone concentration used), we observed diverse antagonistic effects. At the 2-BrEtA highest concentration, a considerable reduction of stomatal closure was observed in WT (from 77% to 10%), while the mutant genotypes were further unresponsive to ABA (5% from 15% in *Atcuaoδ.1* and 9% from 17% in *Atcuaoδ.2*). The other CuAO inhibitor, AG, partially prevented the stomatal closure effects induced by ABA in WT (35% and 24% at 0.1 mM and 1 mM, respectively). However, by itself this inhibitor presented a similar effect on stomatal apertures in all the studied genotypes (10% closure in respect to the Control WT and mutant).

2.3. AtCuAOδ-Driven Production of H_2O_2 Is Involved in the ABA-Induced Stomatal Closure

In order to get insights into the possible role played by the AtCuAOδ-driven production of H_2O_2 in the stomatal closure induced by ABA, WT and *Atcuaoδ* seedlings were treated with the H_2O_2 scavenger N,N^1-dimethylthiourea (DMTU) at the working concentration of 100 μM [26], either alone or in combination with 1, 10, and 100 μM ABA. DMTU reversed the ABA-induced stomatal closure in WT plants (91, 80, and 78%, respectively), whereas it did not significantly affect stomatal aperture under physiological conditions in WT plants, or in 1 and 10 μM ABA-treated and untreated mutants (Figure 2A and Table S1). At the highest ABA concentration used, DMTU reversion was not complete in the mutant genotypes, where a significant closure effect of 8% and 11% was observed in *Atcuaoδ.1* and *Atcuaoδ.2*, respectively. To further investigate the contribution of AtCuAOδ in the ABA-induced H_2O_2 production, reactive oxygen species (ROS) levels in guard cells were analyzed using a chloromethyl derivative of 2′,7′-dichlorodihydrofluorescein diacetate (CM-H_2DCFDA). Figure 3 shows ROS levels through Laser Scanning Confocal Microscopy (LSCM) analysis in both WT and *Atcuaoδ* mutant plants. Under the technical conditions of our analysis, ROS were undetectable in Control WT and Control mutants (Figure 3 left panels), while 100 μM ABA-treatment induced ROS in guard cells of WT plants (Figure 3, upper right panel), as indicated by the green-yellow colored stomata. The signal was absent in *Atcuaoδ.1 and Atcuaoδ.2* mutant seedlings (Figure 3, right lower panels).

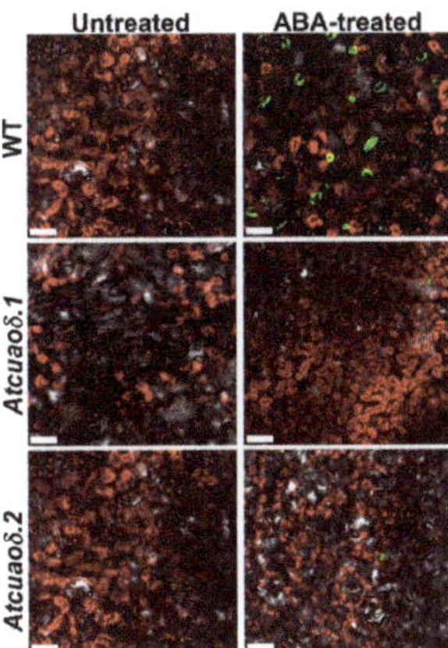

Figure 3. Reactive oxygen species (ROS) levels in guard cells of leaves from 12-day-old seedling. In situ ROS detection in guard cells by Laser Scanning Confocal Microscopy (LSCM) analysis after 2′,7′-dichlorodihydrofluorescein diacetate (CM-H_2DCFDA) staining of leaves from WT, *Atcuaoδ.1*, and *Atcuaoδ.2*, untreated, or 100 μM ABA-treated plants (2 h; green-yellow). Micrographs are representative of those obtained from five independent experiments, each time analyzing leaves from five plants per genotype and treatment. Bar = 50 μm.

2.4. AtCuAOδ Over-Expressing Plants Show Enhanced Stomatal Closure and H_2O_2 Production

The role played by AtCuAOδ-driven production of H_2O_2 in the stomatal closure has been further investigated through the analysis of transgenic Arabidopsis plants over-expressing *AtCuAOδ* (*overAtCuAOδ*). In Figure S3 we show by RT-qPCR and western blot analysis the different levels of the transgene mRNA and protein of the two lines, *overAtCuAOδ* line P9 and *overAtCuAOδ* line P17, used in this work. The latter showed higher expression of the transgenic mRNA and protein than *overAtCuAOδ* line P9 (Figure S3). As shown in Figure 4, both *overAtCuAOδ* lines showed constitutively enhanced stomatal closure (Figure 4A) and similar ROS production in guard cells (Figure 4B) as compared to WT plants.

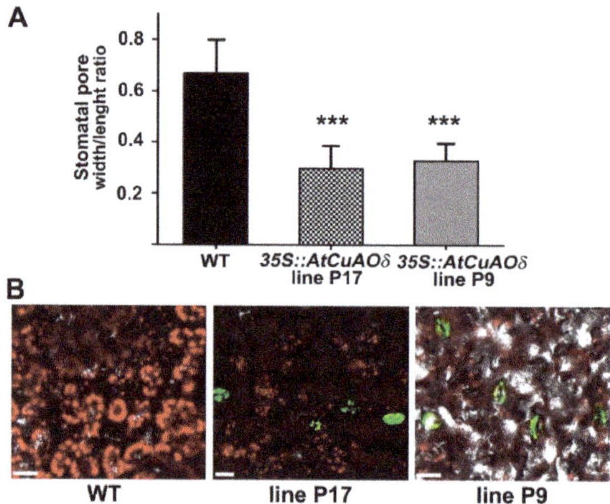

Figure 4. Effect of *AtCuAOδ* over-expression on stomatal pore width/length ratio and ROS levels in guard cells of leaves from 12-day-old WT and *overAtCuAOδ* seedlings. (**A**) Leaves from WT and *overAtCuAOδ* lines P17 and P9 were incubated in the opening solution for 3 h under light to allow stomata opening and then incubated with the fixing solution. Then, the width/length stomatal ratio was measured. Mean values ± SD (*n* = 15) are reported. The significance levels between WT and *overAtCuAOδ* plants are reported. *P* values have been calculated with one-way ANOVA analysis; ***, *p* value equal or less than 0.001. (**B**) In situ ROS detection in guard cells by LSCM analysis after CM-H$_2$DCFDA staining (green-yellow) of leaves from WT and *overAtCuAOδ* lines (P17 and P9). Micrographs are representative of those obtained from five independent experiments, each time analyzing leaves from five plants per genotype and treatment. Bar = 25 μm.

3. Discussion

3.1. AtCuAOδ Plays a Role in the Control of Stomatal Closure in Response to ABA

Our data (Figure 1) show that *AtCuAOδ* gene is regulated by ABA, consistent with the ABA-regulated recognition sites identified in its promoter region. In detail, a maximum of 2.5- to four-fold induction, depending on the ABA concentration used, was observed after 6 h from the onset of treatment. In this regard, it is known that ABA, the water-stress hormonal signal that is considered a valid indicator of water potential status in plants, is involved in defense responses against abiotic stresses, such as drought or high soil saline levels [34–36]. In line with this, several results have suggested that the ABA-responsive AOs [22,25] are likely involved in salt stress responses [37] and in water balance regulation [28].

Furthermore, one of the major roles of this phytohormone is its action on the regulation of stomatal movement in response to variations in water potential [38,39]. Regarding this aspect of plant responses to ABA, our results demonstrate that alterations in the levels of *AtCuAOδ* expression by reverse genetics and over-expression approaches, or AtCuAO enzyme activities by pharmacological treatments with the two known CuAO activity inhibitors, 2-BrEtA or AG [40–42], caused alterations of the hormonal control of guard cells responses. In fact, it was observed a complete to partial unresponsiveness to ABA in stomatal closure with the two homozygous *Atcuaoδ* mutants, compared to the WT (Figure 2A and Table S1), and the lack of ABA responsiveness in the WT in combination treatments involving the two inhibitors (Figure 2B and Table S2). It must be pointed out that this effect was clear even in the case of AG, even if this inhibitor caused by itself a similar stomatal closure effect on all the three studied genotypes (approximately 10% of Control WT and Control mutants). This unspecific effect could be attributed to AG-induced alterations of the plasma membrane potential [43], which could influence stomatal movements. The *overAtCuAOδ* lines instead showed a significantly reduced (50%) stomatal aperture in respect to WT (Figure 4A).

3.2. Vacuolar AtCuAOδ-Dependent H_2O_2 Production Is a Necessary Condition for ABA Regulation of Stomatal Aperture

Figures 3 and 4 show that when compared to the WT, the loss-of-function or the over-expression of the gene encoding the vacuole-resident AtCuAOδ [29] reduced ROS levels of guard cells in ABA-treated mutant genotypes or increased ROS in *overAtCuAOδ* lines. ROS, such as H_2O_2, are ubiquitous metabolites in all aerobic organisms and have been shown to be important signals in many aspects of plant development, including the regulation of stomatal movement [44–46]. Thus, the relation between AtCuAOδ-mediated H_2O_2 production and its involvement in stomatal closure in the ABA transduction pathway is consistent with the role of H_2O_2 in ABA-signaled phenomena.

Interestingly, vacuoles have an important role in the regulation of stomatal pore apertures associated with different environmental or hormonal factors signaling water stress [47–49]. Moreover, ROS can regulate several channel activities located in the tonoplast, which influence ion fluxes, cytosolic pH, and the uptake and release of calcium [30], all of which are involved in modulation of the stomatal aperture. It is, thus, not surprising that an AtCuAO protein identified in the vacuolar proteome [29] can influence stomatal closure as an element in the ABA transduction pathway regulating this phenomenon. No clear contribution of vacuoles in guard cells to the ROS signaling network has been identified [46]. Nevertheless, reports in the literature indicate that vacuoles can be sites of H_2O_2 production [50–52]. Thus, our data might represent a first indication that ROS in the form of H_2O_2 produced by a vacuolar AtCuAO have a physiological role in ABA regulation of stomatal movement.

3.3. Vacuolar AtCuAOδ Cooperates with Different ROS Sources in Regulation of Stomatal Movement

Among CuAO family member AtCuAOδ is not the only CuAO involved in the control of stomatal aperture levels. Indeed, both *AtCuAOβ* and *AtCuAOζ* are expressed in guard cells [25,29], and in the case of *AtCuAOζ* mutant, a reduced ROS level and stomatal closure in response to ABA are observed [25]. This evidence suggests a potential role (*AtCuAOβ*) or effective involvement (*AtCuAOζ*) in the control of the stomatal aperture in Arabidopsis [25,29]. Finally, and in agreement with our data, a CuAO was shown to act in *Vicia faba* during ABA responses involved in the regulation of stomatal apertures through the production of Put-derived H_2O_2 [22], suggesting the existence of a common hormonal response pathway in evolutionarily distant taxa.

The complexity of the ABA signal transduction in guard cells is also highlighted by evidence showing that multiple pathways involving several components and compartments are required in the control of stomatal movements in Arabidopsis. In this context, plasma membrane-located nicotinamide adenine dinucleotide phosphate (NADPH) oxidases (AtrbohD and F [53]), regulated by the ABA-induced phospholipase D (PLDα1 [25]), and ABA-activated OST1 [54], peroxisomal AtCuAOζ [25], and vacuolar AtCuAOδ (in this work) represent multiple ROS sources. These different ROS sources are active in different cellular compartments, targeted by ABA, and necessary for this response, suggesting that a strongly coordinated network is needed for the hormonal control of stomatal closure. Indeed, coordination of ROS signaling from different organelles has been reported for ABA-induced stomatal closure [55]. Furthermore, some evidence shows that ABA presents an apparently minor ROS-independent effect on stomatal closure, as its effects could not be entirely counter-balanced by either ROS scavengers or ROS-biosynthesis inhibitors (Figure 2, Table S1) [25].

Our data revealed that another player is necessary in ABA-mediated regulation of stomatal closure and points out the necessity of understanding the hierarchy of action or eventual synergy of actors involved. The contribution of both NADPH oxidases and AOs has also been proposed in the two phases of ROS production during the hypersensitive response to pathogens, with the former involved in the initial burst or first phase and the latter in the second phase [18,19]. It is possible that a similar mechanism involving both AtCuAOδ and AtCuAOζ, as well as NADPH oxidases, have a role in the ABA control of water balance homeostasis acting through cells or tissues responsible for the regulation of water loss in Arabidopsis. Of note is that an apoplastic PAO and a NADPH oxidase are involved in a feed-forward ROS amplification loop in tobacco, suggesting that both enzymes cooperate in ROS homeostasis in plants [37], while a peroxisomal PAO cross-talks with NADPH oxidase in Arabidopsis to activate mitochondrial alternative oxidase, underlining the complexity of ROS homeostasis and biosynthesis involving different enzymatic systems and subcellular compartments [56]. This picture is further enriched by the downstream effect of NO on the involvement of both NADPH oxidase and AOs in PA-induced stomatal closure in the guard cells of Arabidopsis [57]. The complexity of the events involved in the process of stomatal closure in cases where the increase of ABA levels are a signal of both immediate or prolonged reduced water potential could be explained on the basis of required differential and accurate responses at the cellular or tissue level to the different conditions to which plants are commonly exposed during their growth in-field. Clarification of these mechanisms would be helpful in elucidating potential applications for crop adaptation to changing climate, wherein water stress conditions are becoming increasingly relevant.

4. Materials and Methods

4.1. Plant Materials, Growth Conditions and Treatments

The Columbia-0 (Col-0) ecotype of Arabidopsis was used as WT. The Arabidopsis Col-0 T-DNA insertion lines *Atcuaoδ.1* (SALK_072954.55.00.x line, TAIR accession number 4122972) and *Atcuaoδ.2* (GK-011C04-013046 line, TAIR accession number 4242275) of the gene (At4g12290 TAIR accession number 2139069) were obtained from the Salk Institute Genomic Analysis Laboratory (http://signal.salk.edu/tabout.html; [58]) and from the Bielefeld University CeBiTec/GABI-KAT III [59]. Information on the T-DNA insertion mutants *Atcuaoδ.1* and *Atcuaoδ.2* were obtained respectively from the SIGnAL website (http://signal.salk.edu) and from the GABI-KAT website (https://www.gabi-kat.de/). The Arabidopsis Col-0 transgenic lines constitutively expressing *AtCuAOδ* (*overAtCuAOδ*) were constructed as described below.

Plants were grown in soil or in vitro in a growth chamber at a temperature of 23 °C under long-day conditions (16/8 h photoperiod; 50 μmol m^{-2} s^{-1} and 55% relative humidity). Soil-grown plants were used for identification of the homozygous insertion mutants, for floral dip transformation to prepare over-expression lines, and in all cases for harvesting seeds of the WT and selected lines. For in vitro growth, seeds were surface sterilized as previously described [60]. Seeds were cold stratified at 4 °C and grown in one-half-strength Murashige and Skoog salt mixture supplemented with 0.5% (w/v) sucrose in presence of 0.8% (w/v) agar. In vitro-grown seedlings were used in the analysis of *AtCuAOδ* gene expression, in measurements of stomatal apertures, and in ROS detection.

The analysis of *AtCuAOδ* gene expression upon hormone treatment was performed on Arabidopsis seedlings grown for 12 days on agar medium and then transferred to liquid medium [one-half-strength Murashige and Skoog salt mixture supplemented with 0.5% (w/v) sucrose] containing 1, 10, and 100 μM abscisic acid (ABA; Duchefa) for the described time (0, 1, 3, 6, and 24 h).

Stomatal aperture measurements were performed on 12-day-old Arabidopsis WT plants, *Atcuaoδ* mutants, and *overAtCuAOδ* lines (homozygous T3 generation) grown on agar medium under control conditions in absence of treatment (Control WT, Control *Atcuaoδ.1*, Control *Atcuaoδ.2*, or *overAtCuAOδ* lines). For WT and mutants plants, stomatal aperture levels were also monitored after 2 h treatment with ABA (1, 10, and 100 μM), N,N^1-dimethylthiourea (DMTU; 100 μM), 1, 10, and 100 μM ABA/100 μM DMTU, 2-bromoethylamine (2-BrEtA; 0.5, 5 mM), aminoguanidine (AG; 0.1, 1 mM), 100 μM ABA/0.5 or 5 mM 2-BrEtA, and 100 μM ABA/0.1 or 1 mM AG.

The detection of reactive oxygen species (ROS) in guard cells was analyzed on 12-day-old Arabidopsis WT plants, *Atcuaoδ* mutants, and *overAtCuAOδ* lines grown on agar medium under control conditions. ROS levels in WT and mutant plants were also analyzed after 2 h treatment with 100 μM ABA.

4.2. Identification of the T-DNA Insertional Loss-of-Function Atcuaoδ.1 and Atcuaoδ.2 Mutants

Plants homozygous for the T-DNA insertion were identified by Polymerase Chain Reaction (PCR) on genomic DNA extracted from leaves of soil-grown plants by alkali treatment [61], using gene- and T-DNA-specific primers. *AtCuAOδ* gene-specific primers (*RP-AtCuAOδ/LP-AtCuAOδ*) were designed outside of the 5′ and 3′ ends of the T-DNA insertions and the T-DNA specific primers (*LBa1* for *Atcuaoδ.1*, and *RB1-pAC161* for *Atcuaoδ.2*) were designed at its left border (Figure S1 and Table S3). Due to the proximity of the insertion points in the two mutants, the same *AtCuAOδ* gene-specific primers were used for both the mutants and referred to as *RP-AtCuAOδ/LP-AtCuAOδ* (Figure S1). The genotype of the *Atcuaoδ* mutants was ascertained by two sets of PCR reactions: one using *RP-AtCuAOδ/LBa1* for *Atcuaoδ.1* and *LP-AtCuAOδ/RB1-pAC161* for *Atcuaoδ.2* to determine the presence of the T-DNA insertion and the other using *RP-AtCuAOδ/LP-AtCuAOδ* for both the mutants to verify the absence of the fragment indicative of a WT allele, as the T-DNA insertion originates a non-amplifiable long transcript (Figure S1). The absence of the full-length *AtCuAOδ* gene transcript in *Atcuaoδ.1* and *Atcuaoδ.2* seedlings was analyzed by Reverse Transcription Polymerase Chain Reaction (RT-PCR) of total RNA, using *rtPCR-AtCuAOδ-for1/RP-AtCuAOδ* as gene-specific primers (Figure S1 and Table S3), which would generate in WT an amplicon of 545 bp.

4.3. Construction of the Over-Expressing Transgenic Lines

The transgenic Arabidopsis *overAtCuAOδ* plants were prepared using Gateway technology. The *AtCuAOδ* gene sequence was amplified by PCR from *Arabidopsis* genomic DNA extracted by alkali treatment [61] from agar medium-grown seedlings using the gene-specific primers *overAtCuAOδ-for* and *overAtCuAOδ-rev* (Table S3). The *overAtCuAOδ-rev* primer was designed in order to insert the coding sequence for two Ser residues followed by a 6×His tag prior to the stop codon of the corresponding amplicon. The PCR product was purified and cloned initially into the pDONR 221 vector (Invitrogen), sequenced, and cloned into the pK2GW7 vector [62] through the Gateway recombination system (Invitrogen). The pK2GW7 construct (*35SCaMV::AtCuAOδ-6His*) was checked by sequencing prior to be transferred to *Agrobacterium tumefaciens* (strain GV 301) and then used to transform soil-grown *Arabidopsis* Col-0 WT plants by the floral dip transformation method [63]. Putatively transformed plants were controlled on selective medium (agar medium supplemented with kanamycin at the final concentration of 50 μg/mL) and subsequent PCR analysis of genomic DNA using the gene-specific primer *overAtCuAOδ-for* and a 6×His tag-specific primer (Table S3). Recombinant *AtCuAOδ* expression in *35SCaMV::AtCuAOδ-6His* transgenic plants was determined by RT-qPCR using the gene-specific primers *RTqPCR-AtCuAOδ-for* and the *RTqPCR-AtCuAOδ-rev* (Table S3), as well as western-blot analysis using a rabbit anti-6×His tag antibody conjugated to horseradish peroxidase (Abcam). All the analyses were performed on the third generation (T3) of the lines described herein.

4.4. PCR, RT-PCR and RT-Quantitative PCR (RT-qPCR) Analysis

PCR reactions was carried out with the Dream*Taq*TM DNA Polymerase (Fermentas) in a iCyclerTM ThermalCycler (Bio-Rad) with the following parameters: 2 min of denaturation at 95 °C, 35 cycles of 95 °C for 30 s, 58 °C for 1 min, 72 °C for 1.5 min, and 10 min at 72 °C for the final extension. Total RNA was isolated from 12-day-old whole Arabidopsis seedlings using the RNeasy Plant Mini kit (QIAGEN) following the manufacturer's instructions. DNase digestion was performed during RNA purification using the RNase-Free DNase Set (QIAGEN). For RT-PCR, the first cDNA strand was synthesized from total RNA following the protocol of the ImProm-II Reverse Transcription System (Promega). Ubiquitin-conjugating enzyme 21 (*UBC21*) [64] was used as the internal control to confirm equal amounts of RNA among the various samples, using the primers *UBC21-for* and *UBC21-rev* (Table S3).

RT-qPCR analysis was performed on DNase-treated RNA (4 µg) from 12-day-old whole Arabidopsis seedlings. The cDNA synthesis and PCR amplification were carried out using GoTaq®2-Step RT-qPCR System200 (Promega) according to the manufacturer's protocol. The PCRs were run in a Corbett RG6000 (Corbett Life Science, QIAGEN) utilizing the following program: 95 °C for 2 min, then 40 cycles of 95 °C for 7 s and 60 °C for 40 s. The melting program ramps from 60 °C to 95 °C. rising by 1 °C each step. *AtCuAOδ* specific primers were *RTqPCR-AtCuAOδ-for* and *RTqPCR-AtCuAOδ-rev* (Table S3). *UBC21* (At5g25760) was used as reference gene and specific primers were prepared [64] (*UBC21-For* and *UBC21-Rev*; Table S3). Fold change in the expression of the *AtCuAOδ* was calculated according to the $\Delta\Delta C_q$ method as follows [65], where:

C_q refers to the quantification cycle,

$\Delta C_q = C_{q\ target\text{-}gene} - C_{q\ reference\text{-}gene}$,

$\Delta\Delta C_q = 2^{-[\Delta C_{q\ sample\ time\ X} - \Delta C_{q\ control\text{-}sample\ time\ 0}]}$,

Expression fold-induction = $\Delta\Delta C_{q\ treated\ sample} / \Delta\Delta C_{q\ non\text{-}treated\ sample}$.

The value of ΔC_q at time 0 (control sample) has been assumed to be the reference value for both treated and untreated samples at each experimental time. Accordingly, $\Delta\Delta C_q$ has been calculated as indicated above for both treated and untreated samples at each experimental time. The reported values of expression fold-inductions after treatment are relative to the corresponding expression values of non-treated plants for each time point, with the value for time zero assumed to be one. The software used to control the thermocycler and to analyze data was the Corbett Rotor-Gene 6000 Application Software (version 1.7, Build 87; Corbett Life Science, QIAGEN, Milan, Italy).

4.5. Measurement of Stomatal Aperture

Measurement of stomatal aperture was performed as described previously [66], with slight modifications. In detail, seedlings from 12-day-old Arabidopsis WT plants, *Atcuaoδ* mutants, and *AtCuAOδ* over-expressing lines grown on agar medium were incubated in opening solution (30 mM KCl, 10 mM MES-Tris, pH 6.15) for 3 h under light to allow stomatal opening. Then, seedlings from WT plants and *Atcuaoδ* mutants were incubated for 2 h under light in liquid medium in the absence or presence of ABA 1, 10, and 100 µM, DMTU 100 µM, and ABA 1, 10, and 100 µM/DMTU 100 µM to analyze stomatal aperture.

Treatments with CuAOs inhibitors 2-BrEtA and AG were performed as follows: after 3 h incubation with opening solution, seedlings from WT plants and *Atcuaoδ* mutants were incubated in liquid medium supplemented or not with 2-BrEtA (0.5, 5 mM) or AG (0.1, 1 mM) for 30 min under light, after which ABA at the final concentration of 100 µM was added and further incubated for 2 h under light.

Based on preliminary experiments of ABA dose-response curve, independent experiments were carried out grouping treatments in separate blocks by ABA concentration or CuAO inhibitor type.

Following the various treatments, seedlings from WT plants, *Atcuaoδ* mutants, and *AtCuAOδ* over-expressing lines were treated with a fixing solution (1% glutaraldehyde, 10 mM NaPi pH 7.0, 5 mM $MgCl_2$, and 5 mM EDTA) and incubated for 30 min under light. Images of stomata with the outline of stomatal pores in the focal plane were acquired by a Leica DFC 450C digital camera applied to a Zeiss Axiophot 2 microscope at the magnification of 20×, and stomatal apertures (width/length) were measured using a digital ruler (ImageJ 1.44). Width and length of stomata pores were measured, and stomatal apertures were expressed as the width/length ratio.

4.6. In Situ Detection of Reactive Oxygen Species (ROS) in Guard Cells

ROS production in guard cells was analyzed using a chloromethyl derivative of 2′,7′-dichlorodihydrofluorescein diacetate (CM-H_2DCFDA; Molecular Probes, Invitrogen) as previously described [25,67], with slight modifications. Arabidopsis leaves from 12-day-old seedlings from WT, as well as *AtCuAOδ* insertional mutants and over-expressing lines grown on agar medium, were detached and incubated for 3 h in the assay solution containing 5 mM KCl, 50 µM $CaCl_2$, and 10 mM MES-Tris (pH 6.15), and then 50 µM CM-H_2DCFDA was added to the sample. Leaves were incubated for 30 min at room temperature and then the excess dye was washed out with the fresh assay solution. Collected tissues were again incubated in the assay solution containing 100 µM ABA for 20 min in dark conditions. Images were captured by Laser Scanning Confocal Microscopy (LSCM), using a Leica TCS-SP5 equipped with an Argon laser (Excitation/Emission: ~492–495/517–527 nm) and the Leica Application Suite Advanced Fluorescence (LAS-AF; Leica Microsystems, Milan, Italy).

4.7. Statistics

For RT-qPCR analysis, five independent experiments (in this case representing the biological replicates; $n = 5$) were performed; in each experiment, the seedlings (~200 mg) from four agar plates for every time point and treatment were used. For each cDNA obtained, qPCR was performed in triplicate (technical replicates) and the triplicate mean values have been used in the statistical analysis for each of the five independent experiments.

For the stomatal aperture measurements, three independent experiments were performed for each treatment on the different genotypes. Independent experiments were carried out grouping treatments in separate blocks by ABA concentration or CuAO inhibitor type. A Control and a 100 µM ABA treatment for each of the genotypes analyzed were always included to verify data reproducibility between blocks. For each time, five similarly-sized leaves were harvested from different seedlings for each genotype and treatment. In this case, each of the five leaves from the three experiments was considered a biological replicate for a total of fifteen biological replicates for each genotype and treatment ($n = 15$). For each leaf, four random chosen fields (430 µm × 325 µm) were acquired and approximately 60 stomata were measured, and the mean values were used in the statistical analysis. The results presented are supported by preliminary experiments of ABA dose-response curves. Statistical tests were performed using GraphPad Prism (GraphPad Software) with one-way ANOVA analysis, followed by Sidak's multiple comparison tests. Statistical significance of differences was evaluated by p level, with ns showing not significant, and *, **, ***, and **** p values representing equal or less than 0.05, 0.01, 0.001, and 0.0001, respectively. The LSCM analysis of ROS production by CM-H_2DCFDA-staining was performed on plants from five independent experiments, each of which analyzed leaves from five plants per plant genotype and per treatment, yielding reproducible results. Images from single representative experiments are shown.

Supplementary Materials: The following are available online at http://www.mdpi.com/2223-7747/8/6/183/s1. Figure S1: Characterization of *Atcuaoδ.1* and *δ.2* mutants. Figure S2: Nucleotide and deduced amino acid sequences of *AtCuAOδ* gene (At4g12290, TAIR accession number 2139069) retrieved from TAIR database. Figure S3: Characterization of the lines over-expressing the *AtCuAOδ* gene, showing both protein and mRNA levels and the positive correlation between them. Table S1: Data presented in Figure 2A with the effect of ABA and DMTU on stomata pore width/length ratio. Results from three independent experiments are reported (mean values, SD, and SE). Table S2: Data presented in Figure 2B with the effect of ABA and CuAO inhibitors (2-BrEtA and AG) on stomata pore width/length ratio. Results from three independent experiments are reported (mean values, SD, and SE). Table S3: Primers used in the different PCR procedures indicated in Material and Methods.

Author Contributions: Conceptualization, S.A.G. and A.C. (Alessandra Cona); formal analysis, I.F. and R.A.R.-P.; funding acquisition, R.A., A.C. (Alessandra Cona), and R.A.R.-P.; investigation, I.F., S.A.G., and A.C. (Andrea Carucci); methodology, I.F., S.A.G., and A.C. (Alessandra Cona); supervision, R.A., A.C. (Alessandra Cona), and R.A.R.-P.; validation, I.F.; writing—original draft, S.A.G., A.C. (Andrea Carucci), A.C. (Alessandra Cona), and R.A.R.-P.; writing—review and editing, I.F., P.T., R.A., A.C. (Alessandra Cona), and R.A.R.-P.

Funding: Grant of Excellence Departments, Italian Ministry for University and Research (MIUR-ARTICOLO 1, COMMI 314 – 337 LEGGE 232/2016), is gratefully acknowledged (I.F., S.G., A.C.a., R.A., P.T., and A.C.,);

Ministero dell'Istruzione, dell'Università e della Ricerca (MIUR) Progetti di Ricerca di Interesse Nazionale (PRIN) 2017 (project contract no. 2017ZBBYNC_002 to R.A.) is gratefully acknowledge; Project "Ricerca d'Interesse d'Ateneo-RIA 2015, RIA 2016, RIA 2017, and RIA 2018 from the Università dell'Aquila-Department of Life, Health, and Environmental Sciences are gratefully acknowledge (RAR-P).

Acknowledgments: We thank the Arabidopsis Biological Resource Center for distributing the seeds of the SALK lines. The T-DNA mutant generated in the context of the GABI-Kat program used in this work (GK-011C04-013046) was provided by Bernd Weisshaar (MPI for Plant Breeding Research; Cologne, Germany).

Conflicts of Interest: The authors state that no conflict of interest exists.

References

1. Moschou, P.N.; Wu, J.; Cona, A.; Tavladoraki, P.; Angelini, R.; Roubelakis-Angelakis, K.A. The polyamines and their catabolic products are significant players in the turnover of nitrogenous molecules in plants. *J. Exp. Bot.* **2012**, *63*, 5003–5015. [CrossRef] [PubMed]
2. Tavladoraki, P.; Cona, A.; Federico, R.; Tempera, G.; Viceconte, N.; Saccoccio, S.; Battaglia, V.; Toninello, A.; Agostinelli, E. Polyamine catabolism: Target for antiproliferative therapies in animals and stress tolerance strategies in plants. *Amino Acids* **2012**, *42*, 411–426. [CrossRef] [PubMed]
3. Cona, A.; Rea, G.; Angelini, R.; Federico, R.; Tavladoraki, P. Functions of amine oxidases in plant development and defence. *TRENDS Plant Sci.* **2006**, *11*, 80–88. [CrossRef]
4. Medda, R.; Bellelli, A.; Peč, P.; Federico, R.; Cona, A.; Floris, G. Copper Amine oxidases from plants. In *Copper Amine Oxidases: Structure, Catalytic Mechanism and Role in Pathophysiology*; Floris, G., Mondovì, B., Eds.; Taylor and Francis Group, C.R.C. Press: Boca Raton, FL, USA, 2009; Volume 4, pp. 39–50.
5. Møller, S.G.; McPherson, M.J. Developmental expression and biochemical analysis of the Arabidopsis *ATAO1* gene encoding an H_2O_2-generating diamine oxidase. *Plant J.* **1998**, *13*, 781–791. [CrossRef]
6. Boudart, G.; Jamet, E.; Rossignol, M.; Lafitte, C.; Borderies, G.; Jauneau, A.; Esquerré-Tugayé, M.T.; Pont-Lezica, R. Cell wall proteins in apoplastic fluids of *Arabidopsis thaliana* rosettes: Identification by mass spectrometry and bioinformatics. *Proteomics* **2005**, *5*, 212–221. [CrossRef]
7. Planas-Portell, J.; Gallart, M.; Tiburcio, A.F.; Altabella, T. Copper containing amine oxidases contribute to terminal polyamine oxidation in peroxisomes and apoplast of *Arabidopsis thaliana*. *BMC Plant Biol.* **2013**, *13*, 109. [CrossRef]
8. Naconsie, M.; Kato, K.; Shoji, T.; Hashimoto, T. Molecular evolution of *N*-methylputrescine oxidase in tobacco. *Plant Cell Physiol.* **2014**, *55*, 436–444. [CrossRef] [PubMed]
9. Marina, M.; Maiale, S.J.; Rossi, F.R.; Romero, M.F.; Rivas, E.I.; Gárriz, A.; Ruiz, O.A.; Pieckenstain, F.L. Apoplastic polyamine oxidation plays different roles in local responses of tobacco to infection by the necrotrophic fungus *Sclerotinia sclerotiorum* and the biotrophic bacterium *Pseudomonas viridiflava*. *Plant Physiol.* **2008**, *147*, 2164–2178. [CrossRef]
10. Zarei, A.; Trobacher, C.P.; Cooke, A.R.; Meyers, A.J.; Hall, J.C.; Shelp, B.J. Apple fruit copper amine oxidase isoforms: Peroxisomal MdAO1 prefers diamines as substrates, whereas extracellular $MdAO_2$ exclusively utilizes monoamines. *Plant Cell Physiol.* **2015**, *56*, 137–147. [CrossRef] [PubMed]
11. Pintus, F.; Spanò, D.; Floris, G.; Medda, R. Euphorbia characias Latex Amine Oxidase and Peroxidase: Interacting Enzymes? *Protein J.* **2013**, *32*, 435–441. [CrossRef]
12. Tavladoraki, P.; Cona, A.; Angelini, R. Copper-containing amine oxidases and FAD dependent polyamine oxidases are key players in plant tissue differentiation and organ development. *Front. Plant Sci.* **2016**, *7*, 824. [CrossRef] [PubMed]
13. Cona, A.; Cenci, F.; Cervelli, M.; Federico, R.; Mariottini, P.; Moreno, S.; Angelini, R. Polyamine oxidase, a hydrogen peroxide-producing enzyme, is up-regulated by light and down-regulated by auxin in the outer tissues of the maize mesocotyl. *Plant Physiol.* **2003**, *131*, 803–813. [CrossRef] [PubMed]
14. Tisi, A.; Federico, R.; Moreno, S.; Lucretti, S.; Moschou, P.N.; Roubelakis-Angelakis, K.A.; Angelini, R.; Cona, A. Perturbation of polyamine catabolism can strongly affect root development and xylem differentiation. *Plant Physiol.* **2011**, *157*, 200–215. [CrossRef] [PubMed]
15. Wu, J.; Shang, Z.; Wu, J.; Jiang, X.; Moschou, P.N.; Sun, W.; Roubelakis-Angelakis, K.A.; Zhang, S. Spermidine oxidase-derived H_2O_22 regulates pollen plasma membrane hyperpolarization-activated Ca^{2+}-permeable channels and pollen tube growth. *Plant J.* **2010**, *63*, 1042–1053. [CrossRef] [PubMed]

16. Moschou, P.N.; Paschalidis, K.A.; Delis, I.D.; Andriopoulou, A.H.; Lagiotis, G.D.; Yakoumakis, D.I.; Roubelakis-Angelakis, K.A. Spermidine exodus and oxidation in the apoplast induced by abiotic stress is responsible for H_2O_2 signatures that direct tolerance responses in tobacco. *Plant Cell* **2008**, *20*, 1708–1724. [CrossRef] [PubMed]

17. Angelini, R.; Tisi, A.; Rea, G.; Chen, M.M.; Botta, M.; Federico, R.; Cona, A. Involvement of polyamine oxidase in wound healing. *Plant Physiol.* **2008**, *146*, 162–177. [CrossRef]

18. Yoda, H.; Hiroi, Y.; Sano, H. Polyamine oxidase is one of the key elements for oxidative burst to induce programmed cell death in tobacco cultured cells. *Plant Physiol.* **2006**, *142*, 193–206. [CrossRef] [PubMed]

19. Yoda, H.; Fujimura, K.; Takahashi, H.; Munemura, I.; Uchimiya, H.; Sano, H. Polyamines as a common source of hydrogen peroxide in host- and nonhost hypersensitive response during pathogen infection. *Plant Mol. Biol.* **2009**, *70*, 103–112. [CrossRef]

20. Mitsuya, Y.; Takahashi, Y.; Berberich, T.; Miyazaki, A.; Matsumura, H.; Takahashi, H.; Terauchi, R.; Kusano, T. Spermine signaling plays a significant role in the defense response of *Arabidopsis thaliana* to cucumber mosaic virus. *J. Plant Physiol.* **2009**, *166*, 626–643. [CrossRef]

21. Moschou, P.N.; Sarris, P.F.; Skandalis, N.; Andriopoulou, A.H.; Paschalidis, K.A.; Panopoulos, N.J.; Roubelakis-Angelakis, K.A. Engineered polyamine catabolism preinduces tolerance of tobacco to bacteria and oomycetes. *Plant Physiol.* **2009**, *149*, 1970–1981. [CrossRef]

22. An, Z.; Jing, W.; Liu, Y.; Zhang, W. Hydrogen peroxide generated by copper amine oxidase is involved in abscisic acid-induced stomatal closure in *Vicia faba*. *J. Exp. Bot.* **2008**, *59*, 815–825. [CrossRef]

23. Paschalidis, K.A.; Toumi, I.; Moschou, N.P.; Roubelakis-Angelakis, K.A. ABA-dependent amine oxidases-derived H_2O_2 affects stomata conductance. *Plant Signal. Behav.* **2010**, *5*, 1153–1156.

24. Song, X.G.; She, X.; Yue, P.M.; Liu, Y.E.; Wang, Y.X.; Zhu, X.; Huang, A.X. Involvement of copper amine oxidase (CuAO)-dependent hydrogen peroxide synthesis in ethylene-induced stomatal closure in *Vicia faba*. *Russ. J. Plant Physiol.* **2014**, *61*, 390–396. [CrossRef]

25. Qu, Y.; An, Z.; Zhuang, B.; Jing, W.; Zhang, Q.; Zhang, W. Copper amine oxidase and phospholipase D act independently in abscisic acid (ABA)-induced stomatal closure in *Vicia faba* and *Arabidopsis*. *J. Plant Res.* **2014**, *127*, 533–544. [CrossRef] [PubMed]

26. Ghuge, S.A.; Carucci, A.; Rodrigues Pousada, R.A.; Tisi, A.; Franchi, S.; Tavladoraki, P.; Angelini, R.; Cona, A. The apoplastic copper AMINE OXIDASE1 mediates jasmonic acid-induced protoxylem differentiation in Arabidopsis roots. *Plant Physiol.* **2015**, *168*, 690–707. [CrossRef] [PubMed]

27. Wimalasekera, R.; Tebartz, F.; Scherer, G.F.E. Polyamines, polyamines oxidases and nitric oxide in development, abiotic and biotic stresses. *Plant Sci.* **2011**, *181*, 593–603. [CrossRef]

28. Ghuge, S.A.; Tisi, A.; Carucci, A.; Rodrigues-Pousada, R.A.; Franchi, S.; Tavladoraki, P.; Angelini, R.; Cona, A. Cell wall amine oxidases: New players in root xylem differentiation under stress conditions. *Plants* **2015**, *4*, 489–504. [CrossRef]

29. Ghuge, S.A.; Carucci, A.; Rodrigues-Pousada, R.A.; Tisi, A.; Franchi, S.; Tavladoraki, P.; Angelini, R.; Cona, A. The MeJA-inducible copper amine oxidase AtAO1 is expressed in xylem tissue and guard cells. *Plant Signal. Behav.* **2015**, *10*, e1073872. [CrossRef]

30. Carter, C.; Pan, S.; Zouhar, J.; Avila, E.L.; Girke, T.; Raikhel, N.V. The vegetative vacuole proteome of Arabidopsis thaliana reveals predicted and unexpected proteins. *Plant Cell* **2004**, *16*, 3285–3303. [CrossRef]

31. Bak, G.; Lee, E.J.; Lee, Y.; Kato, M.; Segami, S.; Sze, H.; Maeshima, M.; Hwang, J.U.; Lee, Y. Rapid structural changes and acidification of guard cell vacuoles during stomatal closure require phosphatidylinositol 3,5-bisphosphate. *Plant Cell* **2013**, *25*, 2202–2216. [CrossRef]

32. Winter, D.; Vinegar, B.; Nahal, H.; Ammar, R.; Wilson, G.V.; Provart, N.J. An "Electronic Fluorescent Pictograph" browser for exploring and analyzing large-scale biological data sets. *PLoS ONE* **2007**, *2*, e718. [CrossRef] [PubMed]

33. Swarbreck, D.; Wilks, C.; Lamesch, P.; Berardini, T.Z.; Garcia-Hernandez, M.; Foerster, H.; Li, D.; Meyer, T.; Muller, R.; Ploetz, L.; et al. The Arabidopsis Information Resource (TAIR): Gene structure and function annotation. *Nucleic Acids Res.* **2008**, *36*, D1009–D1014. [CrossRef] [PubMed]

34. Davies, W.; Zhang, J. Root signals and the regulation of growth and development of plants in drying soil. *Annu. Rev. Plant Physiol. Plant Mol. Biol.* **1991**, *42*, 55–76. [CrossRef]

35. Tardieu, F.; Davies, W.J. Stomatal response to abscisic acid is a function of current plant water status. *Plant Physiol.* **1992**, *98*, 540–545. [CrossRef] [PubMed]

36. Yang, Z.; Liu, J.; Tischer, S.V.; Christmann, A.; Windisch, W.; Schnyder, H.; Grill, E. Leveraging abscisic acid receptors for efficient water use in Arabidopsis. *Proc. Natl. Acad. Sci. USA* **2016**, *113*, 6791–6796. [CrossRef] [PubMed]

37. Gémes, K.; Kim, Y.J.; Park, K.Y.; Moschou, P.N.; Andronis, E.; Valassaki, C.; Roussis, A.; Roubelakis-Angelakis, K.A. An NADPH-oxidase/polyamine oxidase feedback loop controls oxidative burst under salinity. *Plant Physiol.* **2016**, *172*, 1418–1431. [CrossRef] [PubMed]

38. Brodribb, T.J.; McAdam, S.A.M. Evolution of the stomatal regulation of plant water content. *Plant Physiol.* **2017**, *174*, 639–649. [CrossRef]

39. Todaka, D.; Zhao, Y.; Yoshida, T.; Kudo, M.; Kidokoro, S.; Mizoi, J.; Kodaira, K.-S.; Takebayashi, Y.; Kojima, M.; Sakakibara, H.; et al. Temporal and spatial changes in gene expression, metabolite accumulation and phytohormone content in rice seedlings grown under drought stress conditions. *Plant J.* **2017**, *90*, 61–78. [CrossRef]

40. Biegański, T.; Osińska, Z.; Maśliński, C. Inhibition of plant and mammalian diamine oxidases by hydrazine and guanidine compounds. *Int. J. Biochem.* **1982**, *14*, 949–953. [CrossRef]

41. Medda, R.; Padiglia, A.; Pedersen, J.Z.; Agro', A.F.; Rotilio, G.; Floris, G. Inhibition of copper amine oxidase by haloamines: A killer product mechanism. *Biochemistry* **1997**, *36*, 2595–2602. [CrossRef]

42. Padiglia, A.; Medda, R.; Pedersen, J.Z.; Lorrap, A.; Peč, P.; Frébort, I.; Floris, G. Inhibitors of plant copper amineoxidases. *J. Enzyme Inhib.* **1998**, *13*, 311–325. [CrossRef] [PubMed]

43. Srivastava, S.K.; Smith, T.A. The effect of some oligoamines and guanidines on membrane permeability in higher plants. *Phytochemistry* **1982**, *21*, 997–1008. [CrossRef]

44. Pei, Z.-M.; Murata, Y.; Benning, G.; Thomine, S.; Klüsener, B.; Allen, G.J.; Grill, E.; Schroeder, J.I. Calcium channels activated by hydrogen peroxide mediate abscisic acid signalling in guard cells. *Nature* **2000**, *406*, 731–734. [CrossRef] [PubMed]

45. Song, Y.; Miao, Y.; Song, C.P. Behind the scenes: The roles of reactive oxygen species in guard cells. *New Phytol.* **2014**, *201*, 1121–1140. [CrossRef] [PubMed]

46. Murata, Y.; Mori, I.C.; Munemasa, S. Diverse stomatal signaling and the signal integration mechanism. *Annu. Rev. Plant Biol.* **2015**, *66*, 369–392. [CrossRef] [PubMed]

47. Peiter, E.; Maathuis, F.J.; Mills, L.N.; Knight, H.; Pelloux, J.; Hetherington, A.M.; Sanders, D. The vacuolar Ca^{2+}-activated channel TPC1 regulates germination and stomatal movement. *Nature* **2005**, *434*, 404–408. [CrossRef]

48. Andrés, Z.; Pérez-Hormaeche, J.; Leidi, E.O.; Schlückingb, K.; Steinhorstb, L.; McLachlanc, D.H.; Schumacherd, K.; Hetheringtonc, A.M.; Kudlab, J.; Cuberoa, B.; et al. Control of vacuolar dynamics and regulation of stomatal aperture by tonoplast potassium uptake. *Proc. Natl. Acad. Sci. USA* **2014**, *111*, E1806–E1814. [CrossRef]

49. Wang, P.C.; Dua, Y.Y.; Hou, Y.J.; Zhao, Y.; Hsu, C.C.; Yuan, F.J.; Zhu, X.H.; Tao, W.A.; Song, C.P.; Zhu, J.K. Nitric oxide negatively regulates abscisic acid signaling in guard cells by S-nitrosylation of OST1. *Proc. Natl. Acad. Sci. USA* **2015**, *112*, 613–618. [CrossRef]

50. Leshem, Y.; Golani, Y.; Kaye, Y.; Levine, A. Reduced expression of the v-SNAREs AtVAMP71/AtVAMP7C gene family in *Arabidopsis* reduces drought tolerance by suppression of abscisic acid-dependent stomatal closure. *J. Exp. Bot.* **2010**, *61*, 2615–2622. [CrossRef]

51. Pradedova, E.V.; Trukhan, I.S.; Nimaeva, O.D.; Salyaev, R.K. Hydrogen peroxide generation in the vacuoles of red beet root cells. *Dokl. Biol. Sci.* **2013**, *449*, 1–4. [CrossRef]

52. Koffler, B.E.; Luschin-Ebengreuth, N.; Stabentheiner, E.; Müller, M.; Zechmann, B. Compartment specific response of antioxidants to drought stress in Arabidopsis. *Plant Sci.* **2014**, *227*, 133–144. [CrossRef] [PubMed]

53. Kwak, J.M.; Mori, I.C.; Pei, Z.M.; Leonhardt, N.; Torres, M.A.; Dangl, J.L.; Bloom, R.E.; Bodde, S.; Jones, J.D.G.; Schroeder, J.I. NADPH oxidase AtrbohD and AtrbohF genes function in ROS-dependent ABA signaling in *Arabidopsis*. *EMBO J.* **2003**, *22*, 2623–2633. [CrossRef] [PubMed]

54. Sirichandra, C.; Wasilewska, A.; Vlad, F.; Valon, C.; Leung, J. The guard cell as a single-cell model towards understanding drought tolerance and abscisic acid action. *J. Exp. Bot.* **2009**, *60*, 1439–1463. [CrossRef] [PubMed]

55. Sierla, M.; Rahikainen, M.; Salojärvi, J.; Kangasjärvi, J.; Kangasjärvi, S. Apoplastic and chloroplastic redox signaling networks in plant stress responses. *Antioxid. Redox Signal.* **2013**, *18*, 2220–2239. [CrossRef] [PubMed]

56. Andronis, E.A.; Moschou, P.N.; Toumi, I.; Roubelakis-Angelakis, K.A. Peroxisomal polyamine oxidase and NADPH-oxidase cross-talk for ROS homeostasis which affects respiration rate in *Arabidopsis thaliana*. *Front. Plant Sci.* **2014**, *5*, 132. [CrossRef] [PubMed]

57. Agurla, S.; Gayatri, G.; Raghavendra, A.S. Polyamines increase nitric oxide and reactive oxygen species in guard cells of Arabidopsis thaliana during stomatal closure. *Protoplasma* **2018**, *255*, 153–162. [CrossRef] [PubMed]

58. Alonso, J.M.; Stepanova, A.N.; Leisse, T.J.; Kim, C.J.; Chen, H.; Shinn, P.; Stevenson, D.K.; Zimmerman, J.; Barajas, P.; Cheuk, R.; et al. Genome-wide insertional mutagenesis of *Arabidopsis thaliana*. *Science* **2003**, *301*, 653–657. [CrossRef]

59. Kleinboelting, N.; Huep, G.; Kloetgen, A.; Viehoever, P.; Weisshaar, B. GABI-Kat SimpleSearch: New features of the *Arabidopsis thaliana* T-DNA mutant database. *Nucleic Acids Res.* **2012**, *40*, D1211–D1215. [CrossRef] [PubMed]

60. Valvekens, D.; Van Montagu, M.; Van Lijsebettens, M. Agrobacterium tumefaciens-mediated transformation of Arabidopsis thaliana root explants by using kanamycin selection. *Proc. Natl. Acad. Sci. USA* **1988**, *85*, 5536–5540. [CrossRef]

61. Klimyuk, V.I.; Carroll, B.J.; Thomas, C.M.; Jones, J.D. Alkali treatment for rapid preparation of plant material for reliable PCR analysis. *Plant J.* **1993**, *3*, 493–494. [CrossRef]

62. Karimi, M.; Inzé, D.; Depicker, A. GATEWAY vectors for Agrobacterium-mediated plant transformation. *Trends Plant Sci.* **2002**, *7*, 193–195. [CrossRef]

63. Clough, S.J.; Bent, A.F. Floral dip: A simplified method for Agrobacterium-mediated transformation of *Arabidopsis thaliana*. *Plant J.* **1998**, *16*, 735–743. [CrossRef] [PubMed]

64. Czechowski, T.; Stitt, M.; Altmann, T.; Udvardi, M.K. Genome-wide identification and testing of superior reference genes for transcript normalization. *Plant Physiol.* **2005**, *139*, 5–17. [CrossRef] [PubMed]

65. Livak, K.J.; Schmittgen, T.D. Analysis of relative gene expression data using real-time quantitative PCR and the $2^{-\Delta\Delta CT}$ Method. *Methods* **2001**, *25*, 402–408. [CrossRef] [PubMed]

66. Jung, C.; Seo, J.S.; Han, S.W.; Koo, Y.J.; Kim, C.H.; Song, S.I.; Nahm, B.H.; Choi, Y.D.; Cheong, J.J. Overexpression of AtMYB44 enhances stomatal closure to confer abiotic stress tolerance in transgenic Arabidopsis. *Plant Physiol.* **2008**, *146*, 623–635. [CrossRef]

67. Munemasa, S.; Hossain, M.; Nakamura, Y.; Mori, I.; Murata, Y. The Arabidopsis calcium-dependent protein kinase, CPK6, functions as a positive regulator of methyl jasmonate signaling in guard cells. *Plant Physiol.* **2011**, *155*, 553–561. [CrossRef]

Review

The Interplay among Polyamines and Nitrogen in Plant Stress Responses

Konstantinos Paschalidis [1,†], Georgios Tsaniklidis [2,†], Bao-Quan Wang [3,†], Costas Delis [4], Emmanouil Trantas [1], Konstantinos Loulakakis [1], Muhammad Makky [5], Panagiotis F. Sarris [6,7,8], Filippos Ververidis [1] and Ji-Hong Liu [9,*]

[1] Department of Agriculture, School of Agricultural Sciences, Hellenic Mediterranean University, Estavromenos, GR-71500 Heraklion, Greece
[2] National Agricultural Research Foundation (NAGREF), GR-71103 Heraklion, Greece
[3] School of Horticulture and Landscape Architecture, Henan Province Engineering Research Center of Horticultural Plant Resource Utilization and Germplasm Enhancement, Henan Institute of Science and Technology, Xinxiang 453003, China
[4] Department of Agriculture, University of the Peloponnese, GR-24100 Kalamata, Greece
[5] Department of Agricultural Engineering, Universitas Andalas, Padang 25163, Indonesia
[6] Biosciences, University of Exeter, Geoffrey Pope Building, Exeter EX4 4QD, UK
[7] Institute of Molecular Biology & Biotechnology, GR-70013 Heraklion, Greece
[8] Department of Biology, University of Crete, GR-70013 Heraklion, Greece
[9] Key Laboratory of Horticultural Plant Biology, College of Horticulture and Forestry Sciences, Huazhong Agricultural University, Wuhan 430070, China
* Correspondence: liujihong@mail.hzau.edu.cn
† These authors contributed equally to this work.

Received: 26 July 2019; Accepted: 28 August 2019; Published: 30 August 2019

Abstract: The interplay between polyamines (PAs) and nitrogen (N) is emerging as a key factor in plant response to abiotic and biotic stresses. The PA/N interplay in plants connects N metabolism, carbon (C) fixation, and secondary metabolism pathways. Glutamate, a pivotal N-containing molecule, is responsible for the biosynthesis of proline (Pro), arginine (Arg) and ornithine (Orn) and constitutes a main common pathway for PAs and C/N assimilation/incorporation implicated in various stresses. PAs and their derivatives are important signaling molecules, as they act largely by protecting and preserving the function/structure of cells in response to stresses. Use of different research approaches, such as generation of transgenic plants with modified intracellular N and PA homeostasis, has helped to elucidate a plethora of PA roles, underpinning their function as a major player in plant stress responses. In this context, a range of transgenic plants over-or under-expressing N/PA metabolic genes has been developed in an effort to decipher their implication in stress signaling. The current review describes how N and PAs regulate plant growth and facilitate crop acclimatization to adverse environments in an attempt to further elucidate the N-PAs interplay against abiotic and biotic stresses, as well as the mechanisms controlling N-PA genes/enzymes and metabolites.

Keywords: polyamines; nitrogen metabolism; abiotic and biotic stress; hydrogen peroxide; antioxidant machinery

---

1. Introduction

Nitrogen metabolites, polyamines (PAs), and several important plant phytohormones, such as ethylene, jasmonates, abscisic acid, salicylic acid, have shown to act as crucial growth regulators that can cross talk with each other in stress signaling processes [1–22]. PA biosynthesis/degradation and their homeostasis undergo extensive alterations in response to various stress conditions, such as cell wall degradation [23] oxidative and developmental stress [2,24–31], phytopathogenic bacteria/fungi/viruses [30], water deficiency [31–33], ammonia toxicity and nutrient availability [34], salinity [2,35–37] and heat [38,39].

Plants absorb nitrogen (N) mostly as nitrate or ammonia ions. The nitrate molecules are enzymically converted to ammonia, which is assimilated in plants for amino acid synthesis. Ammonia assimilation is mainly catalyzed by the glutamine synthetase (GS)/glutamate synthase (GOGAT) cycle [40]. Ammonia detoxification, on the other hand, is catalyzed mainly by glutamate dehydrogenase (GDH); however, under stress conditions GDH partly contributes to ammonia assimilation [34,41]. Furthermore, accumulating evidence shows that stress induces PA export and subsequent oxidations in the apoplast that play a role in production of H_2O_2 [22,30,42]. The apoplastic polyamine oxidase (PAO) in cooperation with the NADPH-oxidase creates a feedforward reactive oxygen species (ROS) magnification loop, affecting the oxidative status and climaxes in programmed cell death (PCD) performance. This loop may be a crucial point in many reactions governing salinity stress resistance, with possible functions spreading outside stress resistance [36]. In tobacco transgenic plants overexpressing *ZmPAO*, we detected greater apoplastic/cytoplasmic contents of H_2O_2 and superoxide, accompanied by increase in antioxidant genes; however, these antioxidants cannot efficiently scavenge ROS [27,28]. On the other hand, repression of *ZmPAO* in young tobacco seedlings enables them to resist short-term salinity, which can be attributed either to higher PA content or to lower ROS contents, because of the perturbed PA apoplastic oxidation [27,28].

The links between PAs and growth-regulatory pathways at molecular, biochemical and physiological levels, suggest that altering the expression of specific PA-response factors could provide a new strategy for targeted PA-response engineering. This review elucidates the concerted roles of N and PAs against plant abiotic and biotic stressors, as well as their interplay mechanisms, as far as the related genes/enzymes and metabolites are concerned, in order to help plants adapt to unfavorable environmental conditions.

2. Major Genes Involved in Abiotic and Biotic Stress Responses

It has been documented that drought and salinity, two main abiotic stress factors, disturb at least 20% of the arable land and nearly 40% of the irrigated land in the world [43,44]. These factors severely limit crop yields and result in the loss of more than US$100 billion per annum to the agricultural sector [45]. Drought represents a reduced soil water capacity, causing a decrease in root water uptake, while salinity leads to enhanced salt ion levels in the soil. Because of both lower water potential in the soil and higher ion levels, higher osmosis in plant cells/tissues might lead to elevated levels of osmolytes in order to achieve osmotic balancing. Furthermore, the higher concentration of ions inside the plant may induce an ionic chain reaction, which increases with the duration of the stress, leading to endocellular permeability of toxic ions. In order to combat with these harsh conditions, the plant can respond by either excluding ions or compartmentalizing them in vacuoles. In this regard, the osmotic phenomenon occurs very quickly and is found in all stress conditions, whereas the ionic phenomenon is quite long, progresses with time, and only under salinity [46]. In comparison to abiotic stress, several biotic stress factors, such as viruses, bacteria, fungi, nematodes, insects, and weeds, cause a direct deprivation of plant nutritional agents, leading to decreased strength in host plants. In agricultural terms, both biotic and abiotic stresses cause dramatical pre- and postharvest damages [47].

Plants adopt specific morphological and cellular alterations by sensing stress signals in order to adapt to environmental conditions. However, very few presumed sensors have been recognized. This is mainly due to the fact that functional genes encoding sensor proteins may exist in a redundant way, so alteration of one gene does not cause substantial phenotype alterations under stress. Alternatively, a sensor protein may be essential to plants and loss-of-function mutants are lethal to plants, prohibiting further analysis [43].

Numerous sets of genes/products are linked to abiotic/biotic stress responses at transcriptional and translational level. Genes leading to effective plant adaptation/tolerance may be categorized into four major groups: (i) Genes coding for enzymes of osmolyte biosynthesis, such as proline, mannitol, glycine, betaine, and trehalose; (ii) genes coding for antioxidant enzymes, such as superoxide dismutase (SOD), peroxidase (POD), and catalase (CAT); (iii) genes coding for stress-induced proteins, such as antifreeze proteins, chaperons, and heat shock proteins; and (iv) genes coding for regulatory proteins, such as protein kinases and transcription factors [44]. So far, a large number of functional and regulatory genes implicated in abiotic and biotic stress responses have been identified in a diversity of plants. In particular, we characterized some functional or regulatory genes controlling metabolism of N, PAs, ethylene, and abscisic acid that are involved in various biological processes [22,33,48–72].

3. Stress-Related Nitrogen Flow and Polyamines

Nitrogen is one of the main essential nutrient elements in plants. The N molecules inside plant cells are derived from soil inorganic N uptake, usually in the form of nitrate and ammonium ions, and from ammonia assimilation, N transport throughout the plant, and N remobilization (Figure 1) [34,73–78].

Nitrogen levels influence plant productivity and quality, due to association with various growth substances that are involved in plant stress responses [31,34,55,79,80]. PAs often increase in plants as a result of N application. They usually result from N-induced higher concentrations of their precursor amino acids, such as Orn and Arg, which are converted to putrescine (Put) [20,31,81]. The interplay among PAs and N is evolving as a major participant in plant stress reactions. The first synthesized amino acid is commonly glutamate (Glu), which participates in N recycling/remobilization into other nitrogenous molecules, ensuring N homeostasis in plants (Figure 1). The Glu, as a central N molecule, leads to biosynthesis of proline (Pro), Orn, Arg, and PAs, which constitute a crucial cooperating pathway for carbon (C) and nitrogen (N) assimilation [78]. Pro, Arg, and Put concentrations in plants are further known as some of the important indicators for both biotic and abiotic stress response [31]. *S*-adenosylmethionine (SAM), which is formed by methionine and involved in ethylene biosynthesis, and Orn, an amino acid involved in the urea cycle, are two important precursor molecules in PA synthesis (Figure 1). Put, Spd, Spm, and thermospermine, in turn, are important products of the organic N, as they are found at relatively high endogenous levels. PAs and their C scaffold are involved in several biochemical pathways. PA catabolism has a crucial role in N/C assimilation/remobilization, as it recycles C and N and produces H_2O_2 by PAO [22,26,39,42,73,78]. However, PA accumulation inside plant cells is the consequence of biosynthesis/catabolism, inter-conversions, and conjugation.

Carbon and N are crucial for developmental and stress responses in plants, in terms of life cycle accomplishment and crop production. Therefore, an appropriate N/C balancing is extremely critical for a variety of physiological and biological processes, including stress response. However, the N/C signaling machineries remain fundamentally undiscovered. The N/C cooperative genome-wide function has revealed that the majority of genes in *Arabidopsis* are over- or under-controlled by C and/or N input [78]. Furthermore, PA remobilization is related with the nitrate transport in parenchymal shoot tissues [82]. PA catabolism, producing H_2O_2 and GABA in the cell wall, is also tightly involved in preserving the N/C homeostasis and balance inside plant tissues [83].

It is widely accepted that the PA/N interplay in plants is of major interest, because it connects N metabolism, C fixation, and secondary metabolism pathways. Stress conditions, such as salinity and drought, increase the activity of proteases causing augmentation of ammonium ions inside cells (Figure 1). Ammonia is converted into glutamine and Glu by GS/GOGAT, respectively. Glu gives Orn that requires higher PA biosynthesis in response to various stressful conditions [41].

Orn is a key amino acid participating in cooperating pathways with major amino acids (Figure 1). Transgenic mouse ornithine decarboxylase (mODC) plants with depleted Orn exhibited higher Glu/Orn conversion into Put, leading to N shortage in the cell and to decreased protein synthesis. The same plants also transformed more Glu into Orn, which was partially compensated for enhanced Glu synthesis from integrated N and C [78]. Therefore, overall N assimilation/partitioning in plants is largely dependent on C availability/reallocation and vice versa. Under threatening abiotic and biotic stress conditions, the plants respond by remobilizing N and C into signaling molecules, such as PAs, Pro, GABA, glycine betaine, and β-Ala [78], as they have stress-protective key roles and partly alleviate ammonia cell toxicity. A Glu-Pro-Arg-PA-GABA coordinated path is therefore of major importance to accomplish an equilibrium among assimilated/partitioned N/C inside plant cells [34,78]. Proteomics and transcriptomics studies on PA-stress interaction and classification of major proteins implicated in important plant developmental/stress responses may provide new insights into the molecular mechanisms underlying these processes [84,85]. Moreover, by RNA-RNA in situ hybridization (ISH) methodologies, we have further elucidated the functions of N/PA genes in crop plants. Use of ISH has helped to identify the localization of PA anabolic and catabolic gene transcripts in tissues, such as the locular parenchyma and the vascular bundles, supporting the viewpoint that Put biosynthetic and catabolic genes are mostly expressed in fast growing tissues and that PAs are strongly implicated in fruit ripening [20].

4. N/PA Biotechnological Approaches for Enhanced Tolerance to Abiotic and Biotic Stress

Plants usually circumvent stress conditions by stimulating appropriate responses that lead to altered metabolism and growth. Tolerance to abiotic stress conditions might be achieved via genetic engineering through modifying the endogenous concentrations of osmoprotectants, by increasing ROS scavenging capacity or by robustly excluding ions with efficient transporter/symporter systems [86]. Taking into account that dissecting the function of stress-related genes would assist in elucidating the potential biochemical and molecular machineries for stress adaptation, enormous efforts and approaches have been expended to unravel the genes/proteins/metabolites associated with a plethora of cellular processes that regulate the complicated character of abiotic and biotic stress resistance [44].

As PAs have pleiotropic roles, their homeostasis control is complex. Genetic transformation of N assimilation/detoxification genes and PA biosynthetic genes coding for GDH, arginine decarboxylase (ADC), ODC, SAM decarboxylase (SAMDC), or Spd synthase (SPDS), significantly enhances abiotic stress resistance in numerous plant species [4,5,34,36–39,41,63,72,87–96].

The GS/GOGAT pathway is the main ammonia assimilation cycle in plants. However, under stress conditions, photorespiratory ammonia may hyperaccumulate due to decreased activity of GOGAT or GS. In this regard, we have discovered alternative metabolic ways like GDH that are activated in order to decrease ammonia buildup in the cells (Figure 1) [34]. Transgenic tobacco plants overexpressing the plant *GDH* gene encoding for the a-subunit polypeptide of GDH (gdh-NAD;A1) also exhibit higher ammonium assimilation activity [41]. Furthermore, transgenic rice overexpressing the *GDH* gene from *Eurotium cheralieri* (a lower organism that has stronger ammonium affinity compared to higher plants) showed higher N assimilation efficacy and yield, especially under low N conditions [87].

Under abiotic stress environment, PAs are apoplastically delivered and oxidized by PAO (Figure 1), generating several intermediates. We revealed two different outputs based on the level of the PA oxidized products. On the one hand, low apoplastic PAO generates less amount of H_2O_2, which in turn initiates a ROS protective pathway that triggers tolerance reactions. On the other hand, high apoplastic PAO could produce a large level of H_2O_2, thus triggering plant cell death (PCD) [22,28,42].

The above scenario illuminates mainly the intercellular PAs' role. In another study, transgenic tobacco plants with down-regulated SAMDC underwent abiotic stress-induced PCD, and displayed lower endocellular levels of soluble Spd and Spm. However, we found that PA contents and apoplastic oxidation in the transgenic plants were unpredictably comparable to those of the wild type [28]. The down-regulated SAMDC transgenics, thus, present a balanced PA interplay among developmental and stress responses [32].

However, during biotic stress an opposite scenario is observed. In the *PAO*-overexpressing tobacco plants, we detected a stress-induced up-regulation of the *PAO* gene upon exposure to infection by *Pseudomonas syringae* pv tabaci [30]. The increased expression of the *ADC* gene may promote PA stability. Spm, in turn, is apoplastically excreted and broken down by the enhanced *PAO*, generating excess H_2O_2 that helps plants to cope with the pathogen attack [30]. Therefore, transgenic plants with increased *PAO* exhibited pre-induced resistance towards infections, including biotrophic and hemibiotrophic diseases [30]. Our stress defense model may represent a pioneering way for creating transgenic plants resistant to both abiotic and biotic stresses.

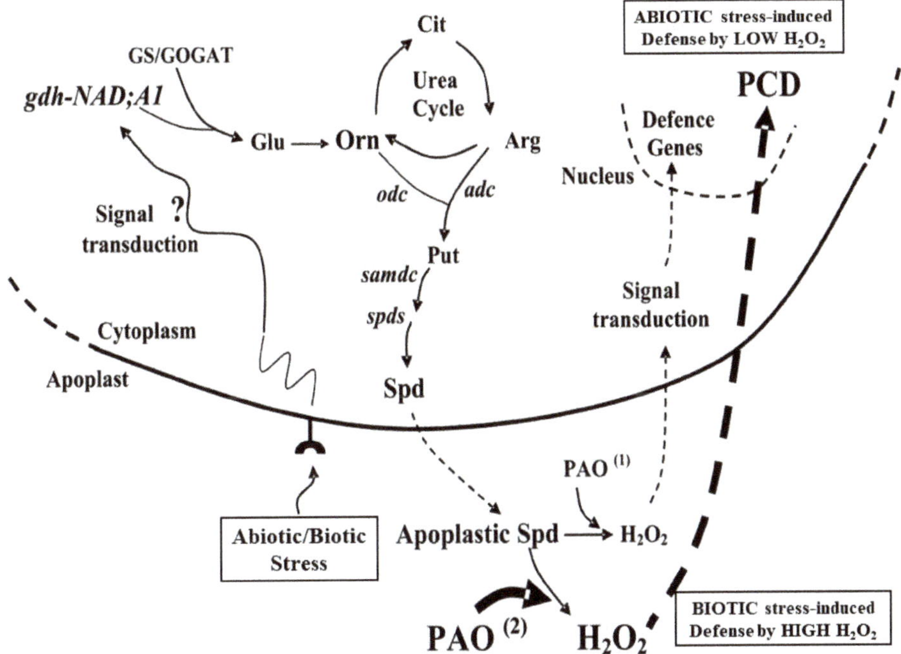

Figure 1. Proposed model for the nitrogen–polyamine (N-PA) interplay in plant abiotic/biotic stress signaling and defense. Abiotic or biotic stress induces proteases activity and increases in ammonium ions inside the cell. The reactive oxygen species (ROS) stress signal triggers the induction of the gene encoding the α-subunit of glutamate dehydrogenase (gdh-NAD; A1), which contributes, together with the GS/GOGAT cycle, to ammonia assimilation [34,41]. Stress-induced glutamate (Glu) production by GDH is diverted to Pro biosynthesis. Stressful conditions cause a further increase in endocellular PAs that are excreted and apoplastically oxidized by polyamine oxidase (PAOs), thus, producing H_2O_2 and numerous N composites. Depending on the level of H_2O_2 produced under abiotic stresses, programmed cell death (PCD; high H_2O_2 levels above a certain threshold) or H_2O_2 scavenging (low levels below a certain threshold) is activated [22,27,28,33]. However, the biotic stress-induced H_2O_2 causes a reverse pattern, as high H_2O_2 levels form an apoplastic "barrier" protecting the plant from fungi and bacteria [30]. Ascorbate peroxidase (APX) and other antioxidant genes are also involved during the protection response. Moreover, PAs are peroxisomally back-converted to generate H_2O_2 and N compounds that could activate Ca^{2+} permeable channels [2,22,28,32,73,97]. PAO [1]: Decrease of PAO activity results in increased Spd and Spm contents and low levels of H_2O_2, leading to expression of defense genes and plant tolerance to abiotic stress, but susceptibility to biotic stress (fungi and bacteria); PAO [2]: Increase of PAO activity results in lower Spd and Spm contents and high levels of H_2O_2, leading to abiotic stress-induced PCD accompanied by plant abiotic stress susceptibility, but tolerance to biotic stress due to high levels of H_2O_2, which form a "barrier" to fungi and bacteria. The other abbreviations are found in the text.

PA catabolism in plants plays a key role in the antioxidant machinery under stress conditions. Overexpression of an apoplastic *PAO* in tobacco plants led to higher expression of antioxidant machinery, including SOD and CAT [27]. However, the induced machinery did not conclude in stress defense, as it represented an effort to neutralize the PAO-produced H_2O_2. Thus, we suggested that continuous PA oxidation may lead to a continuous stress condition. The same genetically modified *Nicotiana tabacum* plants with altered PA/H$_2$O$_2$ levels due to over/underexpression of the *ZmPAO* gene were examined under heat stress. When the *ZmPAO* gene was repressed in transgenic plants, they exhibited better thermotolerance, higher biomass growth, and higher enzymatic and non-enzymatic antioxidant levels. In contrast, the *ZmPAO*-overexpressing plants showed a compromised thermotolerance [38]. Moreover, the *ZmPAO*-underexpressing plants possessed higher Ca^{2+} levels with salinity, associated with lower chlorophyll levels, leaf area and biomass, and a taller phenotype, than the wild type. The *ZmPAO*-overexpressing plants, on the contrary, had a higher number of leaves with slightly greater size and higher antioxidant genes/enzyme levels than the underexpressing ones [37]. Therefore, different phenotypes are found in *PAO*-overexpressing and underexpressing plants under abiotic/biotic stress conditions, revealing a multifaceted character of the apoplastic PAO. In this regard, PAO exerts an important role in rendering plants to survive under both abiotic and biotic stress conditions. It is further proposed that the PAO/NADPH oxidase loop is a focal point in the control of several defense processes in plants, including stress tolerance (Figure 1) [22,24,25,35,36].

5. Conclusion and Perspectives

Metabolic engineering has great potential to enhance abiotic and biotic stress tolerance. Many compounds play dynamic roles in incorporating stress signals, regulating stress response through modifying gene expressions and controlling numerous transporters and biochemical pathways in plants. Modification of a single step in the N/PA cycle (e.g., elevated Glu synthesis via transgenic *GDH* and/or increase of a specific PA via overexpressing of the respective PA biosynthetic gene) might cause a substantial redistribution of the metabolome in the cell. In this regard, for model plants (e.g., *Arabidopsis*) [35,98,99] or plants of industrial use (e.g., grapevine, tobacco, tomato, citrus, etc.) [22,27–30,33,34,37–39,41], metabolic engineering to alter N/PA metabolism and the accompanying N/C accumulation might provide a suitable means for elucidating the physiological mechanisms underlying the increase in crop yield and quality under stress conditions. Extensive knowledge of the N-PA crosstalk by means of engineering technologies may further open new avenues or suggest alternative possibilities for improving the quality of agricultural food products with additional paybacks, such as nutraceuticals and functional components.

Author Contributions: K.P., G.T., B.Q.W., F.V. and J.H.L., collected literatures and wrote the manuscript; C.D., E.T., K.L., M.M. and P.F.S. organized and edited the manuscript; K.P. and J.H.L., proof-read, and finalized the manuscript.

Funding: This work was financially supported by National Key Research and Development Program of China (2018YFD1000300), National Natural Science Foundation of China (31501727, 31320103908), and Hubei Provincial Natural Science Foundation for Innovative Group (2017CFA018).

Acknowledgments: The authors thank anonymous reviewers and the handling editor for providing valuable advice to improve the quality of this review.

Conflicts of Interest: The authors declare no conflicts of interest.

References

1. Ioannidis, N.E.; Malliarakis, D.; Torne, J.M.; Santos, M.; Kotzabasis, K. The over-expression of the plastidial transglutaminase from maize in arabidopsis increases the activation threshold of photoprotection. *Front. Plant Sci.* **2016**, *7*, 635. [CrossRef] [PubMed]

2. Moschou, P.N.; Paschalidis, K.A.; Roubelakis-Angelakis, K.A. Plant polyamine catabolism: The state of the art. *Plant Signal. Behav.* **2008c**, *3*, 1061–1066. [CrossRef] [PubMed]

3. Tavladoraki, P.; Cona, A.; Angelini, R. Copper-containing amine oxidases and fad-dependent polyamine oxidases are key players in plant tissue differentiation and organ development. *Front. Plant Sci.* **2016**, *7*, 824. [CrossRef] [PubMed]

4. Minocha, R.; Majumdar, R.; Minocha, S.C. Polyamines and abiotic stress in plants: A complex relationship. *Front. Plant Sci.* **2014**, *5*, 175. [CrossRef] [PubMed]

5.   Tiburcio, A.F.; Alcazar, R. Potential applications of polyamines in agriculture and plant biotechnology. *Methods Mol. Biol.* **2018**, *1694*, 489–508. [PubMed]

6.   Handa, A.K.; Fatima, T.; Mattoo, A.K. Polyamines: Bio-molecules with diverse functions in plant and human health and disease. *Front. Chem.* **2018**, *6*, 10. [CrossRef]

7.   Fortes, A.M.; Agudelo-Romero, P. Polyamine metabolism in climacteric and non-climacteric fruit ripening. *Methods Mol. Biol.* **2018**, *1694*, 433–447.

8.   Wang, Y.; Ye, X.; Yang, K.; Shi, Z.; Wang, N.; Yang, L.; Chen, J. Characterization, expression, and functional analysis of polyamine oxidases for their role in selenium-induced hydrogen peroxide production in brassica rapa. *J. Sci. Food Agric.* **2019**, *99*, 4082–4093. [CrossRef]

9.   Podlesakova, K.; Ugena, L.; Spichal, L.; Dolezal, K.; De Diego, N. Phytohormones and polyamines regulate plant stress responses by altering gaba pathway. *N. Biotechnol.* **2019**, *48*, 53–65. [CrossRef]

10.   Igarashi, K.; Kashiwagi, K. The functional role of polyamines in eukaryotic cells. *Int. J. Biochem. Cell Biol.* **2019**, *107*, 104–115. [CrossRef]

11.   Bordenave, C.D.; Granados Mendoza, C.; Jimenez Bremont, J.F.; Garriz, A.; Rodriguez, A.A. Defining novel plant polyamine oxidase subfamilies through molecular modeling and sequence analysis. *BMC Evol. Biol.* **2019**, *19*, 28. [CrossRef] [PubMed]

12.   Gimenez, E.; Salinas, M.; Manzano-Agugliaro, F. Worldwide research on plant defense against biotic stresses as improvement for sustainable agriculture. *Sustainability* **2018**, *10*, 391. [CrossRef]

13.   Miao, H.; Sun, P.; Liu, Q.; Liu, J.; Xu, B.; Jin, Z. The agpase family proteins in banana: Genome-wide identification, phylogeny, and expression analyses reveal their involvement in the development, ripening, and abiotic/biotic stress responses. *Int. J. Mol. Sci.* **2017**, *18*, 1581. [CrossRef] [PubMed]

14.   Chen, D.D.; Shao, Q.S.; Yin, L.H.; Younis, A.; Zheng, B.S. Polyamine function in plants: Metabolism, regulation on development, and roles in abiotic stress responses. *Front. Plant Sci.* **2019**, *9*, 1945. [CrossRef] [PubMed]

15.   Xu, J.J.; Liu, T.; Yang, S.C.; Jin, X.Q.; Qu, F.; Huang, N.; Hu, X.H. Polyamines are involved in gaba-regulated salinity-alkalinity stress tolerance in muskmelon. *Environ. Exp. Bot.* **2019**, *164*, 181–189. [CrossRef]

16.   Takahashi, T. Thermospermine: An evolutionarily ancient but functionally new compound in plants. *Methods Mol. Biol.* **2018**, *1694*, 51–59. [PubMed]

17.   Takahashi, T.; Kakehi, J. Polyamines: Ubiquitous polycations with unique roles in growth and stress responses. *Ann. Bot.* **2010**, *105*, 1–6. [CrossRef] [PubMed]

18.   Takano, A.; Kakehi, J.; Takahashi, T. Thermospermine is not a minor polyamine in the plant kingdom. *Plant Cell Physiol.* **2012**, *53*, 606–616. [CrossRef]

19.   Tsaniklidis, G.; Delis, C.; Nikoloudakis, N.; Katinakis, P.; Aivalakis, G. Low temperature storage affects the ascorbic acid metabolism of cherry tomato fruits. *Plant Physiol. Biochem.* **2014**, *84*, 149–157. [CrossRef]

20.   Tsaniklidis, G.; Kotsiras, A.; Tsafouros, A.; Roussos, P.A.; Aivalakis, G.; Katinakis, P.; Delis, C. Spatial and temporal distribution of genes involved in polyamine metabolism during tomato fruit development. *Plant Physiol. Biochem.* **2016**, *100*, 27–36. [CrossRef]

21.   Paschalidis, K.A.; Aziz, A.; Geny, L.; Primikirios, N.I.; Roubelakis-Angelakis, K.A. Polyamines in grapevine. In *Molecular Biology & Biotechnology of the Grapevine*; Roubelakis-Angelakis, K.A., Ed.; Springer: Dordrecht, The Netherlands, 2001; pp. 109–151.

22.   Wang, W.; Paschalidis, K.; Feng, J.C.; Song, J.; Liu, J.H. Polyamine catabolism in plants: A universal process with diverse functions. *Front. Plant Sci.* **2019**, *10*, 561. [CrossRef] [PubMed]

23.   Papadakis, A.K.; Paschalidis, K.A.; Roubelakis-Angelakis, K.A. Biosynthesis profile and endogenous titers of polyamines differ in totipotent and recalcitrant plant protoplasts. *Physiol. Plant* **2005**, *125*, 10–20. [CrossRef]

24.   Papadakis, A.K.; Roubelakis-Angelakis, K.A. Polyamines inhibit nadph oxidase-mediated superoxide generation and putrescine prevents programmed cell death induced by polyamine oxidase-generated hydrogen peroxide. *Planta* **2005**, *220*, 826–837. [CrossRef] [PubMed]

25.   Paschalidis, K.A.; Roubelakis-Angelakis, K.A. Sites and regulation of polyamine catabolism in the tobacco plant. Correlations with cell division/expansion, cell cycle progression, and vascular development. *Plant Physiol.* **2005b**, *138*, 2174–2184. [CrossRef] [PubMed]

26. Paschalidis, K.A.; Roubelakis-Angelakis, K.A. Spatial and temporal distribution of polyamine levels and polyamine anabolism in different organs/tissues of the tobacco plant. Correlations with age, cell division/expansion, and differentiation. *Plant Physiol.* **2005a**, *138*, 142–152. [CrossRef] [PubMed]

27. Moschou, P.N.; Delis, I.D.; Paschalidis, K.A.; Roubelakis-Angelakis, K.A. Transgenic tobacco plants overexpressing polyamine oxidase are not able to cope with oxidative burst generated by abiotic factors. *Physiol. Plant* **2008a**, *133*, 140–156. [CrossRef] [PubMed]

28. Moschou, P.N.; Paschalidis, K.A.; Delis, I.D.; Andriopoulou, A.H.; Lagiotis, G.D.; Yakoumakis, D.I.; Roubelakis-Angelakis, K.A. Spermidine exodus and oxidation in the apoplast induced by abiotic stress is responsible for H2O2 signatures that direct tolerance responses in tobacco. *Plant Cell* **2008b**, *20*, 1708–1724. [CrossRef] [PubMed]

29. Moschou, P.N.; Sanmartin, M.; Andriopoulou, A.H.; Rojo, E.; Sanchez-Serrano, J.J.; Roubelakis-Angelakis, K.A. Bridging the gap between plant and mammalian polyamine catabolism: A novel peroxisomal polyamine oxidase responsible for a full back-conversion pathway in arabidopsis. *Plant Physiol.* **2009b**, *147*, 1845–1857. [CrossRef]

30. Moschou, P.N.; Sarris, P.F.; Skandalis, N.; Andriopoulou, A.H.; Paschalidis, K.A.; Panopoulos, N.J.; Roubelakis-Angelakis, K.A. Engineered polyamine catabolism preinduces tolerance of tobacco to bacteria and oomycetes. *Plant Physiol.* **2009a**, *149*, 1970–1981. [CrossRef]

31. Paschalidis, K.; Moschou, P.N.; Aziz, A.; Toumi, I.; Roubelakis-Angelakis, K.A. Polyamines in grapevine: An update. In *Grapevine Molecular Physiology & Biotechnology*; Roubelakis-Angelakis, K.A., Ed.; Springer: Dordrecht, The Netherlands, 2009a; pp. 207–228.

32. Toumi, I.; Moschou, P.N.; Paschalidis, K.A.; Bouamama, B.; Ben Salem-Fnayou, A.; Ghorbel, A.W.; Mliki, A.; Roubelakis-Angelakis, K.A. Abscisic acid signals reorientation of polyamine metabolism to orchestrate stress responses via the polyamine exodus pathway in grapevine. *J. Plant Physiol.* **2010**, *167*, 519–525. [CrossRef]

33. Paschalidis, K.A.; Toumi, I.; Moschou, P.N.; Roubelakis-Angelakis, K.A. Aba-dependent amine oxidases-derived h2o2 affects stomata conductance. *Plant Signal. Behav.* **2010**, *5*, 1153–1156.

34. Skopelitis, D.S.; Paranychianakis, N.V.; Paschalidis, K.A.; Pliakonis, E.D.; Delis, I.D.; Yakoumakis, D.I.; Kouvarakis, A.; Stephanou, E.; Papadakis, A.K.; Roubelakis-Angelakis, K.A. Abiotic stress generates ros that signal expression of anionic glutamate dehydrogenases to form glutamate for proline synthesis in tobacco and grapevine. *Plant Cell* **2006**, *18*, 2767–2781. [CrossRef] [PubMed]

35. Andronis, E.A.; Moschou, P.N.; Roubelakis-Angelakis, K.A. Peroxisomal polyamine oxidase and nadph-oxidase cross-talk for ros homeostasis which affects respiration rate in arabidopsis thaliana. *Front. Plant Sci.* **2014**, *5*, 132. [CrossRef] [PubMed]

36. Gemes, K.; Kim, Y.J.; Park, K.Y.; Moschou, P.N.; Andronis, E.; Valassaki, C.; Roussis, A.; Roubelakis-Angelakis, K.A. An nadph-oxidase/polyamine oxidase feedback loop controls oxidative burst under salinity. *Plant Physiol.* **2016**, *172*, 1418–1431. [CrossRef] [PubMed]

37. Gemes, K.; Mellidou, I.; Karamanoli, K.; Beris, D.; Park, K.Y.; Matsi, T.; Haralampidis, K.; Constantinidou, H.I.; Roubelakis-Angelakis, K.A. Deregulation of apoplastic polyamine oxidase affects development and salt response of tobacco plants. *J. Plant Physiol.* **2017**, *211*, 1–12. [CrossRef] [PubMed]

38. Mellidou, I.; Karamanoli, K.; Beris, D.; Haralampidis, K.; Constantinidou, H.A.; Roubelakis-Angelakis, K.A. Underexpression of apoplastic polyamine oxidase improves thermotolerance in *Nicotiana tabacum*. *J. Plant Physiol.* **2017**, *218*, 171–174. [CrossRef] [PubMed]

39. Mellidou, I.; Moschou, P.N.; Pankou, C.; Valassakis, C.; Ioannidis, N.; Gémes, K.; Andronis, E.A.; Roussis, A.; Beris, D.; Haralampidis, K.; et al. Nicotiana tabacum plants with silenced s-adenosyl -l-methionine decarboxylase (samdc) reveal a pa-dependent trade-off between growth and tolerance responses. *Front. Plant Sci.* **2016**, *7*, 1–17. [CrossRef] [PubMed]

40. Lea, P.J.; Miflin, B.J. Alternative route for nitrogen assimilation in higher plants. *Nat. Rev. Genet.* **1974**, *251*, 614–616. [CrossRef]

41. Skopelitis, D.S.; Paranychianakis, N.V.; Kouvarakis, A.; Spyros, A.; Stephanou, E.G.; Roubelakis-Angelakis, K.A. The isoenzyme 7 of tobacco nad(h)-dependent glutamate dehydrogenase exhibits high deaminating and low aminating activities in vivo. *Plant Physiol.* **2007**, *145*, 1726–1734. [CrossRef]

42. Moschou, P.N.; Roubelakis-Angelakis, K.A. Polyamines and programmed cell death. *J. Exp. Bot.* **2014**, *65*, 1285–1296. [CrossRef]

43. Zhu, J.-K. Abiotic stress signaling and responses in plants. *Cell* **2016**, *167*, 313–324. [CrossRef] [PubMed]

44. Singhal, P.; Jan, A.T.; Azam, M.; Haq, Q.M.R. Plant abiotic stress: A prospective strategy of exploiting promoters as alternative to overcome the escalating burden. *Front. Life Sci.* **2016**, *9*, 52–63. [CrossRef]

45. Shabala, S.; Bose, J.; Fuglsang, A.T.; Pottosin, I. On a quest for stress tolerance genes: Membrane transporters in sensing and adapting to hostile soils. *J. Exp. Bot.* **2016**, *67*, 1015–1031. [CrossRef] [PubMed]

46. Kosová, K.; Vítámvás, P.; Urban, M.O.; Prášil, I.T.; Renaut, J. Plant abiotic stress proteomics: The major factors determining alterations in cellular proteome. *Front. Plant Sci.* **2018**, *9*, 1–22. [CrossRef]

47. Romero, F.M.; Maiale, S.J.; Rossi, F.R.; Marina, M.; Ruiz, O.A.; Garriz, A. Polyamine metabolism responses to biotic and abiotic stress. *Methods Mol. Biol.* **2018**, *1694*, 37–49. [PubMed]

48. Gago, C.; Drosou, V.; Paschalidis, K.; Guerreiro, A.; Miguel, G.; Antunes, D.; Hilioti, Z. Targeted gene disruption coupled with metabolic screen approach to uncover the leafy cotyledon1-like4 (ll14) function in tomato fruit metabolism. *Plant Cell Rep.* **2017**, *36*, 1065–1082. [CrossRef] [PubMed]

49. Manganaris, G.A.; Drogoudi, P.; Goulas, V.; Tanou, G.; Georgiadou, E.C.; Pantelidis, G.E.; Paschalidis, K.A.; Fotopoulos, V.; Manganaris, A. Deciphering the interplay among genotype, maturity stage and low-temperature storage on phytochemical composition and transcript levels of enzymatic antioxidants in prunus persica fruit. *Plant Physiol. Biochem.* **2017**, *119*, 189–199. [CrossRef] [PubMed]

50. Dhima, K.; Vasilakoglou, I.; Stefanou, S.; Gatsis, T.; Paschalidis, K.; Aggelopoulos, S.; Eleftherohorinos, I. Differential competitive and allelopathic ability of cyperus rotundus on solanum lycopersicum, solanum melongena and capsicum annuum. *Arch. Agron. Soil Sci.* **2016**, *62*, 1250–1263. [CrossRef]

51. Dhima, K.; Vasilakoglou, I.; Paschalidis, K.A.; Gatsis, T.; Keco, R. Productivity and phytotoxicity of six sunflower hybrids and their residues effects on rotated lentil and ivy-leaved speedwell. *Field Crop. Res.* **2012**, *136*, 42–51. [CrossRef]

52. Goumenaki, E.; Karidis, Z.; Paschalidis, K.A. Assessment of tropospheric ozone impact on crops in crete (greece) using snap bean as bioindicator. *Acta Hortic.* **2012**, *938*, 401–407. [CrossRef]

53. Makky, M.; Paschalidis, K.A.; Dhima, K.; Manganaris, A. A new rapid gas chromatographic method for ethylene, respirational, and senescent gaseous production of climacteric fruits stored in prolonged low temperature. *Proc. Int. Conf. Agric. Environ. Biol. Sci. (AEBS-2014)* **2014a**, *1*, 24–25.

54. Makky, M.; Paschalidis, K.A.; Dhima, K.; Mangganaris, A. Tomato fruits (solanaceae lycopersicon esculentum mill.) feedback mechanism in the presence of exogenous ethylene under prolonged chilling temperature storage. *J. Nutr. Pharm. Res.* **2015**, *1*, 4–12.

55. Ninou, E.G.; Paschalidis, K.A.; Mylonas, I.G.; Vasilikiotis, C.; Mavromatis, A.G. The effect of genetic variation and nitrogen fertilization on productive characters of greek oregano. *Acta Agric. Scand. Sect. B—Soil Plant Sci.* **2017b**, *67*, 372–379. [CrossRef]

56. Vasilakoglou, I.; Dhima, K.; Paschalidis, K.; Gatsis, T.; Zacharis, K.; Galanis, M. Field bindweed (convolvulus arvensis l.) and redroot pigweed (amaranthus retroflexus l.) control in potato by pre-or post-emergence applied flumioxazin and sulfosulfuron. *Chil. J. Agric. Res.* **2013**, *73*, 24–30. [CrossRef]

57. Mantzorou, A.; Navakoudis, E.; Paschalidis, K.; Ververidis, F. Microalgae: A potential tool for remediating aquatic environments from toxic metals. *Int. J. Environ. Sci. Technol.* **2018**, *15*, 1815–1830. [CrossRef]

58. Paschalidis, K.A.; Moschou, P.N.; Toumi, I.; Roubelakis-Angelakis, K.A. Polyamine anabolic/catabolic regulation along the woody grapevine plant axis. *J. Plant Physiol.* **2009b**, *166*, 1508–1519. [CrossRef]

59. Mougiou, N.; Trikka, F.; Trantas, E.; Ververidis, F.; Makris, A.; Argiriou, A.; Vlachonasios, K.E. Expression of hydroxytyrosol and oleuropein biosynthetic genes are correlated with metabolite accumulation during fruit development in olive, olea europaea, cv. Koroneiki. *Plant Physiol. Biochem.* **2018**, *128*, 41–49. [CrossRef]

60. Papadakis, I.E.; Tsiantas, P.I.; Tsaniklidis, G.; Landi, M.; Psychoyou, M.; Fasseas, C. Changes in sugar metabolism associated to stem bark thickening partially assist young tissues of eriobotrya japonica seedlings under boron stress. *J Plant Physiol.* **2018**, *231*, 337–345. [CrossRef]

61. Trantas, A.E.; Sarris, P.F.; Mpalantinaki, E.; Papadimitriou, M.; Ververidis, F.; Goumas, D.E. First report of xanthomonas hortorum pv. Hederae causing bacterial leaf spot on ivy in greece. *Plant Disease* **2016**, *100*, 1–10. [CrossRef]

62. Trantas, E.A.; Mpalantinaki, E.; Pagoulatou, M.; Sarris, P.F.; Ververidis, F.; Goumas, D.E. First report of bacterial apical necrosis of mango caused by pseudomonas syringae pv. Syringae in greece. *Plant Disease* **2017**, *101*, 1541. [CrossRef]

63. Wang, W.; Wu, H.; Liu, J.H. Genome-wide identification and expression profiling of copper-containing amine oxidase genes in sweet orange (citrus sinensis). *Tree Genet. Genomes* **2017**, *13*, 31. [CrossRef]

64. Wu, H.; Fu, B.; Sun, P.; Xiao, C.; Liu, J.H. A nac transcription factor represses putrescine biosynthesis and affects drought tolerance. *Plant Physiol.* **2016**, *172*, 1532–1547. [CrossRef] [PubMed]

65. Gong, X.; Zhang, J.; Hu, J.; Wang, W.; Wu, H.; Zhang, Q.; Liu, J.H. Fcwrky70, a wrky protein of fortunella crassifolia, functions in drought tolerance and modulates putrescine synthesis by regulating arginine decarboxylase gene. *Plant Cell Environ.* **2015**, *38*, 2248–2262. [CrossRef] [PubMed]

66. Huang, X.S.; Zhang, Q.; Zhu, D.; Fu, X.; Wang, M.; Zhang, Q.; Moriguchi, T.; Liu, J.H. Ice1 of poncirus trifoliata functions in cold tolerance by modulating polyamine levels through interacting with arginine decarboxylase. *J. Exp. Bot.* **2015**, *66*, 3259–3274. [CrossRef] [PubMed]

67. Wang, J.; Liu, J.H.; Yu, A.; Xiang, Y.; Kurosawa, T.; Nada, K.; Ban, Y. Cloning, biochemical identification, and expression analysis of a gene encoding s-adenosylmethionine decarboxylase in citrus sinensis osbeck. *J. Hortic. Sci. Biotechnol.* **2010**, *85*, 219–226. [CrossRef]

68. Fu, X.Z.; Chen, C.W.; Wang, Y.; Liu, J.H.; Moriguchi, T. Ectopic expression of mdspds1 in sweet orange (citrus sinensis osbeck) reduces canker susceptibility: Involvement of H_2O_2 production and transcriptional alteration. *BMC Plant Biol.* **2011**, *11*, 55. [CrossRef] [PubMed]

69. Shi, J.; Fu, X.Z.; Peng, T.; Huang, X.S.; Fan, Q.J.; Liu, J.H. Spermine pretreatment confers dehydration tolerance of citrus in vitro plants via modulation of antioxidative capacity and stomatal response. *Tree Physiol.* **2010**, *30*, 914–922. [CrossRef] [PubMed]

70. Zhang, Q.; Wang, M.; Hu, J.; Wang, W.; Fu, X.; Liu, J.H. Ptrabf of poncirus trifoliata functions in dehydration tolerance by reducing stomatal density and maintaining reactive oxygen species homeostasis. *J. Exp. Bot.* **2015**, *66*, 5911–5927. [CrossRef]

71. Zhang, X.; Wang, W.; Wang, M.; Zhang, H.Y.; Liu, J.H. The mir396b of poncirus trifoliata functions in cold tolerance by regulating acc oxidase gene expression and modulating ethylene-polyamine homeostasis. *Plant Cell Physiol.* **2016**, *57*, 1865–1878. [CrossRef]

72. Liu, J.-H.; Wang, W.; Wu, H.; Gong, X.; Moriguchi, T. Polyamines function in stress tolerance: From synthesis to regulation. *Front. Plant Sci.* **2015**, *6*, 1–10. [CrossRef]

73. Moschou, P.N.; Wu, J.; Cona, A.; Tavladoraki, P.; Angelini, R.; Roubelakis-Angelakis, K.A. The polyamines and their catabolic products are significant players in the nitrogenous turnover in plants. *J. Exp. Bot.* **2012**, *63*, 5003–5015. [CrossRef] [PubMed]

74. Loulakakis, K.A.; Primikirios, N.I.; Nikolantonakis, M.A.; Roubelakis-Angelakis, K.A. Immunocharacterization of vitis vinifera l. Ferredoxin-dependent glutamate synthase, and its spatial and temporal changes during leaf development. *Planta* **2002**, *215*, 630–638. [CrossRef] [PubMed]

75. Syntichaki, K.M.; Loulakakis, K.A.; Roubelakis-Angelakis, K.A. The amino-acid sequence similarity of plant glutamate dehydrogenase to the extremophilic archaeal enzyme conforms to its stress-related function. *Gene* **1996**, *168*, 87–92. [CrossRef]

76. Loulakakis, K.A.; Roubelakis-Angelakis, K.A.; Kanellis, A.K. Regulation of glutamate dehydrogenase and glutamine synthetase in avocado fruit during development and ripening. *Plant Physiol.* **1994**, *106*, 217–222. [CrossRef] [PubMed]

77. Loulakakis, K.A.; Roubelakis-Angelakis, K.A. Plant nad(h)-glutamate dehydrogenase consists of two subunit polypeptides and their participation in the seven isoenzymes occurs in an ordered ratio. *Plant Physiol.* **1991**, *97*, 104–111. [CrossRef]

78. Majumdar, R.; Barchi, B.; Turlapati, S.A.; Gagne, M.; Minocha, R.; Long, S.; Minocha, S.C. Glutamate, ornithine, arginine, proline, and polyamine metabolic interactions: The pathway is regulated at the post-transcriptional level. *Front. Plant Sci.* **2016**, *7*, 78. [CrossRef]

79. Ninou, E.; Paschalidis, K.; Mylonas, I. Essential oil responses to water stress in greek oregano populations. *J. Essent. Oil Bear. Plants* **2017a**, *20*, 12–23. [CrossRef]

80. Makky, M.; Paschalidis, K.A.; Dhima, K.; Manganaris, A. Harnessing untapped bio-ethylene sources from tomatoes climacteric effluent. *Proc. Int. Conf. Agric. Environ. Biol. Sci. (AEBS-2014)* **2014b**, *1*, 24–25.

81. Serapiglia, M.J.; Minocha, R.; Minocha, S.C. Changes in polyamines, inorganic ions and glutamine synthetase activity in response to nitrogen availability and form in red spruce (picea rubens). *Tree Physiol.* **2008**, *28*, 1793–1803. [CrossRef]

82. Tong, W.; Imai, A.; Tabata, R.; Shigenobu, S.; Yamaguchi, K.; Yamada, M.; Hasebe, M.; Sawa, S.; Motose, H.; Takahashi, T. Polyamine resistance is increased by mutations in a nitrate transporter gene nrt1.3 (atnpf6.4) in arabidopsis thaliana. *Front. Plant Sci.* **2016**, *7*, 1–10. [CrossRef]

83. Wu, Q.Y.; Ma, S.Z.; Zhang, W.W.; Yao, K.B.; Chen, L.; Zhao, F.; Zhuang, Y.Q. Accumulating pathways of gamma-aminobutyric acid during anaerobic and aerobic sequential incubations in fresh tea leaves. *Food Chem.* **2018**, *240*, 1081–1086. [CrossRef] [PubMed]

84. Liu, Y.; Liang, H.; Lv, X.; Liu, D.; Wen, X.; Liao, Y. Effect of polyamines on the grain filling of wheat under drought stress. *Plant Physiol. Biochem.* **2016**, *100*, 113–129. [CrossRef] [PubMed]

85. Tanou, G.; Ziogas, V.; Belghazi, M.; Christou, A.; Filippou, P.; Job, D.; Fotopoulos, V.; Molassiotis, A. Polyamines reprogram oxidative and nitrosative status and the proteome of citrus plants exposed to salinity stress. *Plant Cell Environ.* **2014**, *37*, 864–885. [CrossRef] [PubMed]

86. Nguyen, H.C.; Lin, K.H.; Ho, S.L.; Chiang, C.M.; Yang, C.M. Enhancing the abiotic stress tolerance of plants: From chemical treatment to biotechnological approaches. *Physiol. Plant* **2018**, *164*, 452–466. [CrossRef] [PubMed]

87. Tang, D.; Peng, Y.; Lin, J.; Du, C.; Yang, Y.; Wang, D.; Liu, C.; Yan, L.; Zhao, X.; Li, X.; et al. Ectopic expression of fungal ecgdh improves nitrogen assimilation and grain yield in rice. *J. Integr. Plant Biol.* **2018**, *60*, 85–88. [CrossRef] [PubMed]

88. Liu, J.H.; Ban, Y.; Wen, X.P.; Nakajima, I.; Moriguchi, T. Molecular cloning and expression analysis of an arginine decarboxylase gene from peach (prunus persica). *Gene* **2009**, *429*, 10–17. [CrossRef] [PubMed]

89. Liu, J.H.; Moriguchi, T. Changes in free polyamine titers and expression of polyamine biosynthetic genes during growth of peach in vitro callus. *Plant Cell Rep.* **2007**, *26*, 125–131. [CrossRef] [PubMed]

90. Liu, J.H.; Nada, K.; Honda, C.; Kitashiba, H.; Wen, X.P.; Pang, X.M.; Moriguchi, T. Polyamine biosynthesis of apple callus under salt stress: Importance of the arginine decarboxylase pathway in stress response. *J. Exp. Bot.* **2006**, *57*, 2589–2599. [CrossRef]

91. Liu, J.H.; Nakajima, I.; Moriguchi, T. Effects of salt and osmotic stresses on free polyamine content and expression of polyamine biosynthetic genes in vitis vinifera. *Biol. Plant.* **2011**, *55*, 340–344. [CrossRef]

92. Sun, P.; Zhu, X.; Huang, X.; Liu, J.H. Overexpression of a stress-responsive myb transcription factor of poncirus trifoliata confers enhanced dehydration tolerance and increases polyamine biosynthesis. *Plant Physiol. Biochem.* **2014**, *78*, 71–79. [CrossRef]

93. Wang, B.Q.; Zhang, Q.F.; Liu, J.H.; Li, G.H. Overexpression of ptadc confers enhanced dehydration and drought tolerance in transgenic tobacco and tomato: Effect on ros elimination. *Biochem. Biophys. Res. Commun.* **2011b**, *413*, 10–16. [CrossRef] [PubMed]

94. Wang, J.; Sun, P.P.; Chen, C.L.; Wang, Y.; Fu, X.Z.; Liu, J.H. An arginine decarboxylase gene ptadc from poncirus trifoliata confers abiotic stress tolerance and promotes primary root growth in arabidopsis. *J. Exp. Bot.* **2011a**, *62*, 2899–2914. [CrossRef] [PubMed]

95. Wang, W.; Liu, J.H. Genome-wide identification and expression analysis of the polyamine oxidase gene family in sweet orange (citrus sinensis). *Gene* **2015**, *555*, 421–429. [CrossRef] [PubMed]

96. Wang, W.; Liu, J.H. Cspao4 of citrus sinensis functions in polyamine terminal catabolism and inhibits plant growth under salt stress. *Sci. Rep.* **2016**, *6*, 31384. [CrossRef] [PubMed]

97. Wu, J.; Qu, H.; Shang, Z.; Jiang, X.; Moschou, P.N.; Roubelakis-Angelakis, K.A.; Zhang, S. Spermidine oxidase-derived H2O2 activates downstream ca2+ channels which signal pollen tube growth in pyruspyrifolia. *Plant J.* **2010**, *63*, 1042–1053. [CrossRef] [PubMed]

Plants **2019**, *8*, 315

98. Zarza, X.; Atanasov, K.E.; Marco, F.; Arbona, V.; Carrasco, P.; Kopka, J.; Fotopoulos, V.; Munnik, T.; Gomez-Cadenas, A.; Tiburcio, A.F.; et al. Polyamine oxidase 5 loss-of-function mutations in arabidopsis thaliana trigger metabolic and transcriptional reprogramming and promote salt stress tolerance. *Plant Cell Environ.* **2017**, *40*, 527–542. [CrossRef]

99. Fincato, P.; Moschou, P.N.; Ahou, A.; Angelini, R.; Roubelakis-Angelakis, K.A.; Federico, R.; Tavladoraki, P. The members of arabidopsis thaliana pao gene family exhibit distinct tissue- and organ-specific expression pattern during seedling growth and flower development. *Amino Acids* **2012**, *42*, 831–841. [CrossRef] [PubMed]

Article

Fruit Architecture in Polyamine-Rich Tomato Germplasm Is Determined via a Medley of Cell Cycle, Cell Expansion, and Fruit Shape Genes

Raheel Anwar [1,2]**, Shazia Fatima** [1]**, Autar K. Mattoo** [3] **and Avtar K. Handa** [1,*]

[1] Department of Horticulture and Landscape Architecture, 625 Agriculture Mall Drive, Purdue University, West Lafayette, IN 47906, USA; raheelanwar@uaf.edu.pk (R.A.); shaziafatima66@gmail.com (S.F.)

[2] Institute of Horticultural Sciences, University of Agriculture, Faisalabad, Punjab 38040, Pakistan

[3] Sustainable Agricultural Systems Laboratory, U.S. Department of Agriculture, Agricultural Research Service, The Henry A. Wallace Beltsville Agricultural Research Center, Beltsville, MD 20705, USA; autar.mattoo@usda.gov

* Correspondence: ahanda@purdue.edu; Tel.: +1-765-494-1339

Received: 25 August 2019; Accepted: 24 September 2019; Published: 29 September 2019

Abstract: Shape and size are important features of fruits. Studies using tomatoes expressing *yeast Spermidine Synthase* under either a constitutive or a fruit-ripening promoter showed obovoid fruit phenotype compared to spherical fruit in controls, suggesting that polyamines (PAs) have a role in fruit shape. The obovoid fruit pericarp exhibited decreased cell layers and pericarp thickness compared to wild-type fruit. Transgenic floral buds and ovaries accumulated higher levels of free PAs, with the bound form of PAs being predominant. Transcripts of the fruit shape genes, *SUN1* and *OVATE*, and those of *CDKB2*, *CYCB2*, *KRP1* and *WEE1* genes increased significantly in the transgenic ovaries 2 and 5 days after pollination (DAP). The levels of cell expansion genes *CCS52A/B* increased at 10 and 20 DAP in the transgenic fruits and exhibited negative correlation with free or bound forms of PAs. In addition, the cell layers and pericarp thickness of the transgenic fruits were inversely associated with free or bound PAs in 10 and 20 DAP transgenic ovaries. Collectively, these results provide evidence for a linkage between PA homeostasis and expression patterns of fruit shape, cell division, and cell expansion genes during early fruit development, and suggest role(s) of PAs in tomato fruit architecture.

Keywords: putrescine; spermidine; spermine; tomato; spermidine synthase; fruit shape; cell division; cell expansion

---

1. Introduction

Domestication of tomato has led to different phenotypes, including diversity in fruit shape, color, and size [1]. QTL mapping and genomic analyses have identified several loci underlying the observed diversity in shape and size of tomato fruit [1–5]. Genes that regulate fruit architecture include shape and size genes, namely, *CNR/FW2.2* [6], *OVATE* [2], *FAS* [7], *SUN1* [8], *fw11.3* [9], *LC* [10,11], *fs8.1* [12], and *KLUH/FW3.2* [13]. The *FW2.2* locus encodes a protein with homology to a cell-membrane-localized ras-like G-protein, which has been implicated in controlling ~47% of variation in fruit mass in *Solanum pimpinellifolium* and *S. pennellii* [6,14]. A mutation in the *FW2.2* promoter inhibits cell division during flower development and causes a larger fruit phenotype [15]. Another gene *SUN1* encodes a protein harboring an IQ67 domain and affects cell number along the entire proximal-distal axis, resulting in fruit elongation [16–18]. *OVATE* family proteins (OFPs) and TONNEAU1 Recruiting Motif proteins affect fruit shape by regulating cell division patterns during ovary development [19,20]. It is known that a mutation in the carboxyl-terminal domain of *OVATE* results in changing fruit shape from round-to pear-shaped [2]. *SUN* has a stronger effect on transcriptome than *OVATE* and *fs8.1*. Auxin has been implicated in regulating the expression of some of the fruit shape genes [18], but little is known about other molecules that may affect fruit size and shape.

Polyamines (PAs)—putrescine (Put), spermidine (Spd), and spermine (Spm), are ubiquitous polycations in all organisms, including plants, implicated in a myriad of developmental and physiological processes, including cell proliferation [21–24]. In plants, PAs play roles in biotic and abiotic stresses [25]; fate of flower, fruit, and seed development [26]; leaf and flower senescence [27,28]; in vitro somatic embryogenesis and organogenesis [29,30]; and fruit ripening and shelf life [27,31]. PAs homeostasis is a genetically regulated process with tight PAs homeostasis in most organisms and their perturbation results in altered phenotypes [26]. Exogenous application of PAs was found to increase the expression of cell division genes *CYCA* and *CYCB* in tobacco BY-2 cell cultures [32]. Overexpression of *KRP1*, a cyclin-dependent kinase inhibitor 1, affected cell division and led to several altered phenotypes in plants, including flower morphology and plant size [33].

PAs are predominantly present during growth and development in plants, in particular during cell division and elongation in apical shoots and meristems before flowering [21,34,35]. A crosstalk among various PAs and plant hormones has been implicated in several growth and developmental processes [36]. Transgenic tomato fruits expressing *CDKA1* under the control of a fruit-specific promoter increased cell division, resulting in higher thickness of pericarp and placenta, and larger septa and columella [37]. We previously developed isogenic tomato lines homozygous for expression of yeast Spd synthase (*ySpdSyn*) with altered levels of Spd and Spm in developing and ripening tomato fruit [27]. These isogenic transgenic fruits with altered PA homeostasis resulted in architectural changes in fruit such as a more obovoid phenotype, compared to an otherwise spherical phenotype in wild-type (WT) fruit. The altered phenotype of transgenic fruit manifests during early fruit development and is associated with higher levels of Spd/Spm. Our results show strong positive correlations between transcript levels of fruit shape-regulating genes (*SUN1* and *OVATE*), cell cycle genes (*CDKB2*, *CYCB2*, *KRP1* and *CCS52B*), and endogenous levels of free and bound PAs. These findings provide first molecular evidence for the role of PAs as fruit architecture regulators responsible for the obovoid phenotype of tomato fruit.

2. Results

2.1. Transgenic Expression of ySpdSyn Caused Architectural Changes in Tomato Fruit

The isogenic tomato germplasm homozygous for transgene *ySpdSyn* under the control of either constitutive CaMV35S (lines C4 and C15) or fruit-specific SlE8 promoters (line E8-8) have been previously described [27]. Fruits from the *ySpdSyn* transgenic lines exhibited an obovoid shape, compared to the relatively spherical phenotype of parental Wild-type cv. Ohio 8245 fruits (Figure 1a,b). The altered phenotype transition—from more spherical to more obovoid—of the transgenic fruit occurs at early stages of fruit development and continues thereafter to full maturity (Figure 1a,b). The effect of transgene was more pronounced under the constitutive CaMV35S promoter than under the developmentally regulated fruit-ripening SlE8 promoter. Collectively, these results associate the expression of *ySpdSyn* transgene with the observed altered tomato fruit architecture from round to obovoid.

The cytological examination of pericarp cell layers, cell size, and cell thickness of developing ovaries of WT and the three transgenic lines was performed at 5 days before pollination (DBP) and 5, 10 and 20 days after pollination (DAP) to ascertain the nature of the structural alterations. The medial-lateral sections of the fruit pericarp tissue from WT and transgenic lines at 10 and 20 DAP were stained with 0.04% toluidine blue–O, and are shown in Figure 2a,b; quantified data for the pericarp thickness and cell layer number from −5 to 20 DAP are shown in Figure 2c,d, respectively. The thickness of pericarp in all the three independent transgenic lines, particularly C15 and E8-8 fruit, was significantly reduced compared to WT fruit (Figure 2c), which was associated with a decreased number of cell layers at 20 DAP (Figure 2d). On an average, fruit pericarp of WT fruit tissues at 20 DAP had 38 cell layers while the transgenic fruit pericarp from C4, C15, and E8-8 lines had 32, 28, and 28 cell layers, respectively (Figure 2d). The cell sizes in endocarp, mesocarp, and exocarp of the transgenic fruit were reduced compared to WT fruits, with significant reduction in the cell size of the 10 DAP mesocarp of C4 and C15 lines, as compared to WT fruit mesocarp (Figure 2e). At 20 DAP, however, the mesocarp cell size was unchanged in fruit of C4 and C15 lines, but significantly decreased in E8-8 fruit compared to WT fruit (Figure 2f). The cell sizes in 20 DAP fruit increased by 2- to 11-fold in endocarp, mesocarp, and exocarp, compared to the 10 DAP fruit, implicating a shift from cell division to cell expansion mode (Figure 2g). This increase was much larger in endocarp and mesocarp of the transgenic C4 and C15 fruits compared to WT fruit at 20 DAP.

Figure 1. Morphometric changes in transgenic tomato fruit expressing *ySpdSyn*. Phenotype (**a**) and fruit shape index (**b**) of field grown wild-type (WT, wild-type cv. Ohio 8245) and transgenic fruits expressing *ySpdSyn* under CaMV35S (C4 and C15) and fruit specific E8 (E8-8) promoters. The fruits shown represent average growth and development stage of indicated genotype. The white line in the bottom left corner (Figure 1a) represents 10 mm on the original scale. Vertically cut tomato fruits were scanned and analyzed with tomato analyzer 3.0 software (Figure 1b). Fruit shape index is length-to-width ratio of the fruit. Error bars in Figure 1b represent standard error of means (n > 3 biological replicates, where each replicate had at least 50 tomato fruits). Abbreviations: MG or G—mature green stage; BR or B—breaker stage; Ri or R—red ripe stage of tomato fruit development.

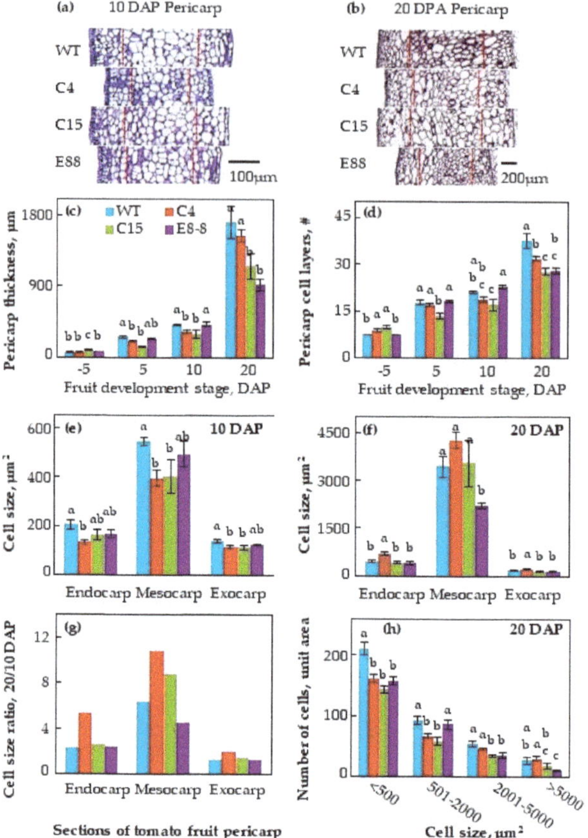

Figure 2. Histological analysis of WT and transgenic fruitlets at 5 days before pollination and at 5, 10 and 20 days after pollination (DAP). Toluidine blue–O staining of WT and transgenic fruitlets at (**a**) 10 DAP and (**b**) 20 DAP. (**c**) Changes in pericarp thickness; (**d**) number of anticlinal cell layers in pericarp; (**e**) cell size at 10 DAP and (**f**) 20 DAP; (**g**) cell size ratio of 20 DAP/10 DAP in endocarp (single innermost cell layer), mesocarp (middle 50% of the pericarp), and exocarp (2 outer cell layers) of tomato ovaries; (**h**) number of cells in each category of cell area within each genotype. Flowers were tagged and ovaries from flowers at 5 days before pollination and at 5, 10, and 20 DAP were fixed in 100% methanol, vertically sectioned and stained with 0.04% toluidine blue–O. Digital images of pericarp sections were acquired using AperioScan and analyzed using ImageScope 11. Average cell size (**e,f**) was calculated by dividing the total number of cells with the area of endocarp, mesocarp, or exocarp. Shown are average ± standard error (n ≥ 3 biological replicates). Different letters above the standard error bars indicate significant difference (at 95% confidence interval) among genotypes within the pericarp section.

In order to determine the phenotypic basis of reduced pericarp thickness of the fruit from transgenic lines, we evaluated the distribution of smaller and larger cells in mediolateral pericarp slices. The pericarp of the transgenic fruits from C4, C15, and E8-8 lines had a reduced number of cells per unit area than the WT pericarp (Figure 2h). Reduction in cell size was seen in almost all cell types, small to large. However, the percent of total distribution of small and large cells ranging from ≤500 to ≥5000 remained similar in all the genotypes (data not shown).

2.2. Expression of ySpdSyn Transgene and SlSpdSyn in Floral Buds and Fertilized Ovaries

The expression of *ySpdSyn* in floral buds and fertilized ovaries of the transgenic and WT tomato lines was quantified by qRT-PCR to evaluate the association between transgene expression and altered phenotype. The *CaMV35S:ySpdSyn* transgene was expressed in C4 tissues as early as 5 DBP and in the fertilized ovaries at 2 to 20 DAP. The transgene expression in C4 was much higher in 5 DAP ovaries than the other stages of floral buds and ovaries examined (Figure 3a). Transcripts of *E8:ySpdSyn* transgene were also detectable at 5 DBP and 5 DAP stages, but the levels were much lower than the *CaMV35S:ySpdSyn* transcripts at all the fruit developmental stages examined (Figure 3a inset). Expression of the endogenous *SlSpdSyn* gene was also regulated during the floral bud development, with high levels observed in 5 DBP and 10 DAP fruits (Figure 3b). Other stages had low levels of *SlSpdSyn* transcripts (Figure 3b). Transgene under the CaMV35 promoter increased *SlSpdSyn* transcripts at 5 DBP and that under SlE8 at 10 DAP stages of fruit development (Figure 3b). Constitutive expression of *ySpdSyn* in C15 line exhibited upregulated expression of endogenous *SlSpdSyn* gene to as much as 5.8-fold in 10 DAP ovaries when compared with WT fruit tissues (Table S1).

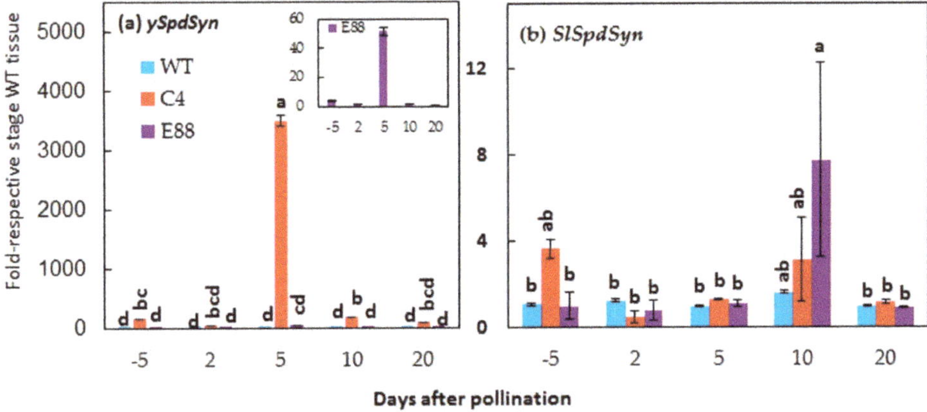

Figure 3. Changes in steady state levels of (**a**) *ySpdSyn* and (**b**) *SlSpdSyn* transcripts during early development of WT and *ySpdSyn* expressing tomato fruits. Total RNA from three biological replicates of floral buds at 5 days before pollination, and flower ovaries at 2, 5, 10, and 20 days after pollination were independently extracted and reverse transcribed. The levels of *ySpdSyn* and *SlSpdSyn* transcripts were determined using qRT-PCR with gene-specific primers (Table S3). The inset in upper panel shows the transcript levels of *E8:ySpdSyn* transgene in E8-8 tissues at a higher magnification. The relative expression levels were calculated by the $2^{-\Delta\Delta C_T}$ method using *SlACTIN* (Solyc04g011500.2.1) as housekeeping gene and plotted as fold-respective to WT tissues. Shown are average ± standard error. Statistical analyses were performed using the XLSTAT ANOVA method using −5 DAP WT as reference.

2.3. Expression Patterns of Fruit Size and Shape Genes in Transgenic Fruits

The expression patterns of *SUN1*, *OVATE*, and *FW2.2* in isogenic transgenic fruit are shown as fold change to corresponding WT fruit at 5 DBP and 2, 5, 10, and 20 DAP, respectively, in Figure 4. Insets in Figure 4 show the expression patterns for *SUN1*, *OVATE*, and *FW2.2* in WT floral buds and developing ovaries as fold-*ACTIN* (a house keeping gene) transcript levels. In WT floral tissues, the levels of *SUN1* and *OVATE* transcripts were upregulated at 5/20 DAP and 2 DAP, respectively, whereas *FW2.2* transcripts significantly increased at 10 and 20 DAP during floral and ovary development (Figure 4 inset). The patterns of *OVATE* and *SUN1* transcript levels in Ohio 8245 (WT) fruit is consistent with the WT fruit associated with the non-mutated *OVATE* gene [2]. Similar results were obtained in C15 fruits (Table S1).

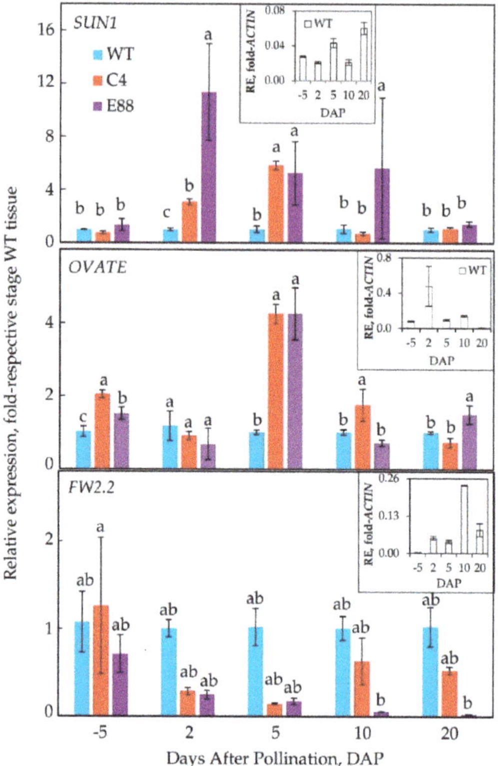

Figure 4. Changes in transcript levels of fruit shape-related genes *SUN1*, *OVATE*, and *FW2.2* in WT and *ySpdSyn*-expressing transgenic tomato floral buds and flower ovaries. Transcripts were quantified using qRT-PCR and relative expression levels were calculated by the $2^{-\Delta\Delta C_T}$ method using *SlACTIN* (Solyc04g011500.2.1) as housekeeping gene and plotted as fold-respective to WT tissues. Other details are the same as in the legend in Figure 3 legend.

A single base pair mutation from GAA to TAA in *OVATE* gene has been linked with functionality of this gene [2]. We checked the nucleotide sequence of *OVATE* in tomato cultivar Ohio 8245 and its isogenic transgenic lines C4, C15, and E8-8 to determine if WT or a mutated gene was residing in under-study genotypes. *OVATE* transcripts cloned from WT and three transgenic lines fruits showed absence of any mutation and thus would have more spherical shape phenotypes as observed in this investigation (Figure S1). Taken together, these results suggested that the observed obovoid fruit phenotype of transgenic line was associated with *ySpdSyn* transgene and not with the *OVATE* gene.

Patterns of transcript levels of the three fruit shape genes in transgenic lines (C4, C15 and E8-8) varied significantly compared to WT fruit (Figure 4; Table S1). *SUN1* transcripts were at higher levels in E8-8 than C4 or WT fruits at most stages of ovary development (Figure 4). The levels of *SUN1* transcripts in C4 were higher than WT at 2 DAP and 5 DAP stages but declined in 20 DAP ovaries in all the transgenic lines compared to WT fruit ovaries (Figure 4; Table S1). The *OVATE* transcript levels were significantly higher (>4-fold) in 5 DAP ovaries from C4 and E8-8 lines compared to WT ovaries. At other stages, the patterns of *OVATE* transcripts in transgenic ovaries were variable compared to the WT ovaries, with significantly higher levels at −5 and 10 DAP in C4 ovaries and at −5 and 20 DAP in E8-8 ovaries (Figure 4; Table S1). Consistent significant patterns for *FW2.2* transcripts levels in the fruits from C4 and E8-8 were not obtained, but generally, they were lower at all stages of ovary development (Figure 4, Table S1).

2.4. Expression Patterns of Selected Cell Division and Cell Expansion Genes in ySpdSyn Transgenic and Wild-Type Tomato

The relative steady state transcript levels of several genes implicated in regulating cell division and cell expansion were examined in floral buds and ovaries from WT, C4, and E8-8 lines. These included cyclin (*CYC*), cyclin-dependent kinases (*CDKs*), *FSM1* (inhibitor of cell expansion), and *CCS52A/CCS52B* (promoters of cell expansion) and their interacting partners *KPR1* and CDK inhibitor *WEE1*. In WT and transgenic flower and ovary tissues, the *CDKA1* transcript levels did not change significantly from 5 DBP to 20 DAP (Figure 5, Figure S2). In WT tissues, the *CDKB2* transcripts significantly decreased from 5 DBP to 2 DAP, followed by several-fold increase until 10 DAP before a precipitous decline at 20 DAP during WT fruit development (Figure S2). The two transgenic lines showed different pattern compared to WT tissues with the *CDKB2* transcripts being about 6-fold higher in C4 and E8-8 ovaries at 2 DAP compared to WT. Thereafter, a noticeable decrease in these genes occurred in all genotypes (Figure 5). In C15 tissues, the *CDKB2* transcripts significantly increased several-fold from 5 DBP to 2 DAP before precipitously declining at 20 DAP (Table S1).

Expression patterns for the three cyclins, *CYCA2*, *CYCB2*, *CYCD3* (one from each of the three cyclin genes families), were different. Overall, their transcript levels were in the increasing order of *CYCB2* > *CYCD3* > *CYCA2* (Figure 5, Figure S2). Levels of the *CYCA2* transcripts in WT ovaries were variable but not significantly different at any stage of floral and ovary development (Figure S2). In contrast, the transcript levels of *CYCB2* and *CYCD3* in WT tissues were about 4- and 12-fold higher, respectively, at 10 DAP before perceptible decline by 20 DAP (Figure S2). These transcript patterns in WT fruit are similar to that reported previously. The patterns of three cyclin gene transcripts in all transgenic ovaries were different. In the E8-8 tissues, several-fold increase in the *CYCA2* transcripts at 2 and 10 DAP stages was apparent, but this was not seen in C4 tissues (Figure 5). The steady state level of *CYCB2* was about 11-fold higher in both the transgenic lines compared to WT tissues at 2 DAP before declining notably in the transgenic tissues (Figure 5). The *CYCD3* transcripts levels had a bimodal pattern in E8-8 tissues showing about a 6-fold increase at 2 DAP and 5 DAP, while a 6-fold increase was obtained in C4 tissue only at 2 DAP (Figure 5). The steady state levels of all cyclin genes examined decreased considerably in 20 DAP tissues in all genotypes.

The pattern of *KRP1* (a CDK inhibitor) transcript accumulation did not change significantly during the floral and ovary development of WT (Figure S2). The C4 ovaries had 7-fold higher accumulation at 2 DAP, but in E8-8 ovaries, there was no change in *KRP1* transcripts at 2 DAP (Figure 5). The levels of CDK inhibitor gene, *WEE1*, increased until 10 DAP in WT (Figure S2), whereas in C4 and E8-8 tissues 6- to 7-fold increase in *WEE1* transcript levels was apparent at 2 DAP (Figure 5).

The transcript levels of *FSM1*, a cell expansion gene, increased several-fold in WT ovaries from 2 to 10 DAP and declined thereafter (Figure S2), a pattern similar to that reported in the facultative parthenocarpic line L-179 (*pat-2/pat-2*) [38]. The *FSM1* transcript levels in the C4 line were highest at 5 DBP (floral buds) and in E8-8 at 20 DAP compared to the other stages of development (Figure 5). The cell expansion promoting genes, *CCS52A* and *CCS52B*, had unique and differing expression patterns among the three genotypes analyzed (Figure 5). In WT tissues, transcript level of *CCS52A* was highest at 10 and 20 DAP and that of *CCS52B* was highest at 5 DBP and 10 DAP (Figure S2). The *CCS52A* transcript levels were significantly high at 5 DBP in C4 and E8-8, and at 5 DAP only in E8-8 ovaries (Figure 5). Highest levels of *CCS52B* transcripts during flower and ovary development was observed in E8-8 at 20 DAP (Figure 5).

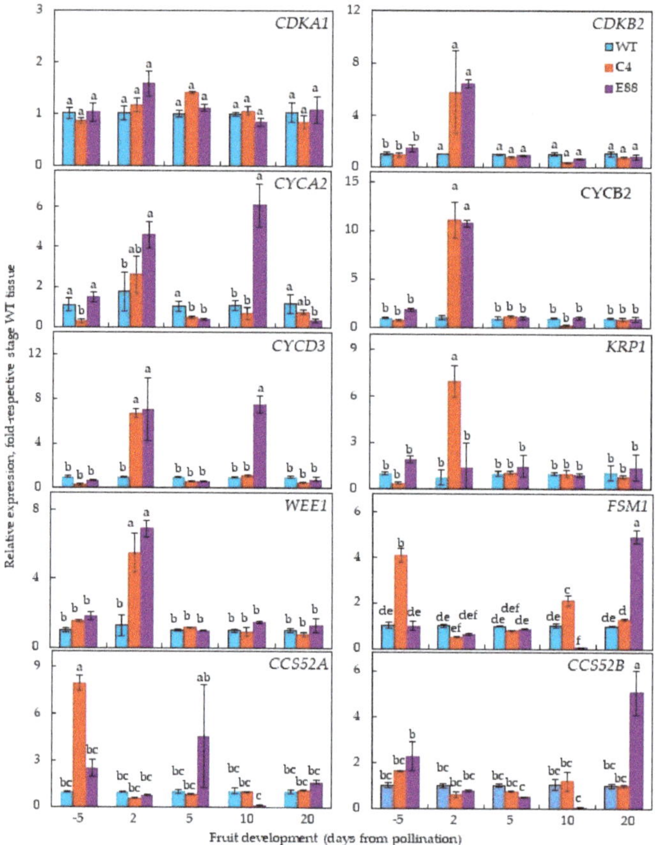

Figure 5. Changes in steady state transcript levels of cyclin-dependent kinases (*CDKA1* and *CDKB2*), cyclins (*CYCA2*, *CYCB2* and *CYCD3*), CDK1-inhibitors (*KRP1* and *WEE1*), and cell expansion-regulating genes (*FSM1*, *CCS52A* and *CCS52B*) in WT and *ySpdSyn*-expressing transgenic tomato floral buds and flower ovaries. Transcripts were quantified using qRT-PCR and relative expression levels were calculated by the $2^{-\Delta\Delta C_T}$ method using *SlACTIN* (Solyc04g011500.2.1) as housekeeping gene and plotted as fold-respective WT tissues. Other details were the same as described in the Figure 3 legend.

2.5. Levels of Free and Bound PAs in Floral Buds and Developing Ovaries

The free and bound levels of Put, Spd, and Spm were quantified in floral tissues of WT and the transgenic lines (C4, C15, E8-8) at −10, −5, 2, 5, 10, and 20 DAP (Figure 6, Table S2). Put, Spd, and Spm levels varied in all the three genotypes examined, with transgenic fruit having greater increase in bound forms of these PAs (Figure 6; Table S2). In the WT, the highest levels of free PAs were found at −5 and 5 DAP for Put, 5 and 10 DAP for Spd, and 5 DAP for Spm. The levels of free PAs declined dramatically at 20 DAP stage of WT fruit. In the transgenic lines, free Put levels were lower than the WT at all the developmental stages examined, except for the E8-8 fruit, which had significantly higher free Put level at 5 DBP than the other stages of development (Figure 6; Table S2). In contrast, levels of bound Put in transgenic lines increased several-fold from 10 DBP to 5 DAP and then decreased to levels comparable with WT at 20 DAP (Figure 6).

Free Spd levels in the WT line slighty increased from 10 DBP to 10 DAP stages and then declined at 20 DAP. For WT tissues, the content of bound Spd was several-fold lower than free Spd at all the fruit developmental stages (Figure 6). Compared to the WT, the free Spd levels had similar pattern of accumulation in developing floral buds and ovaries from the transgenic lines. The C4 and E8-8 fruit had the highest levels of free Spd at 5 DAP and 5 DBP, respectively (Figure 6). Remarkably, the bound Spd levels were high in all the transgenic flowers and developing ovaries at most stages of development from 10 DBP to 5 DAP, as compared to the WT line, but declined after 10 DAP in all the transgenic lines. The bound Spd levels peaked at 5 DBP in all three transgenic lines i.e., C4, C15, and E8-8 (Figure 6; Table S2).

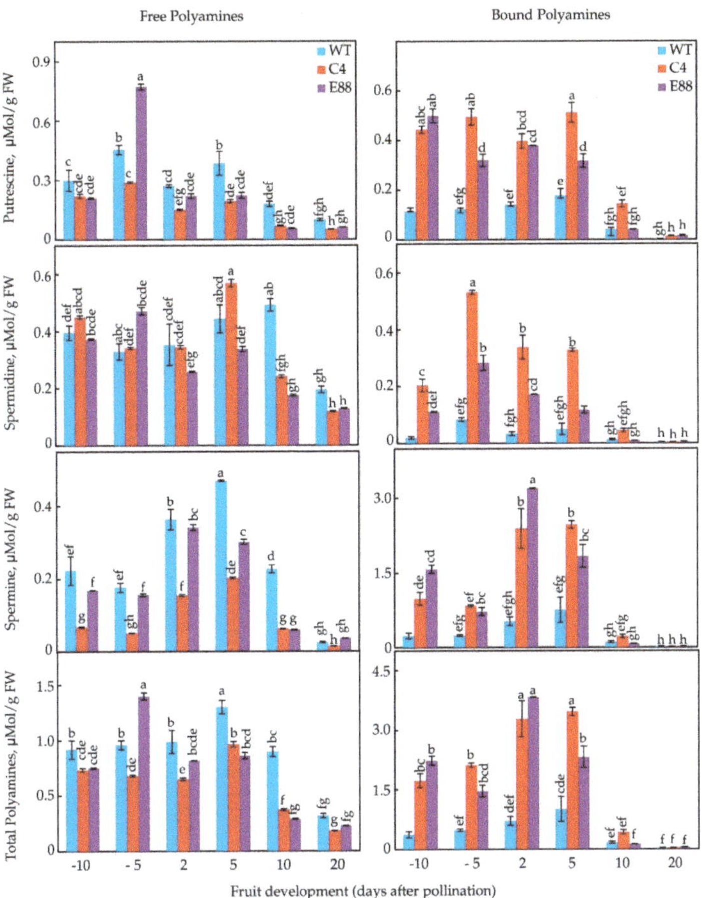

Figure 6. Changes in the levels of free and bound forms of putrescine, spermidine, and spermine (nmol/g FW) in floral buds and ovaries of WT and *ySpdSyn*-expressing transgenic tomato plants. Free and bound PAs were extracted and quantified by HPLC, as described in the Material and Methods section. Shown are average ± standard error (n ≥ 3 biological replicates). Similar letters above standard error bars indicate non-significant difference (at 95% confidence interval) among genotypes.

The levels of free Spm in WT floral tissues were lower than its bound form throughout the development of floral and ovary tissues (Figure 6). Levels of free Spm increased from 5 DBP to 5 DAP before declining in both C4 and E8-8 genotypes (Figure 6). Like Spd, bound form of Spm in C4 and E8-8 was several-fold higher than the free form throughout the development of floral and ovary tissues. The pattern of bound form was similar in the three transgenic lines (Figure 6; Table S2). The highest bound Spm levels were found in E8-8 ovaries at 2 DAP (Figure 6). The C15 and C4 tissues had similar patterns (Table S2).

The levels of bound forms of Put, Spd, and Spm in WT tissues ranged from 2% to 52%, 4% to 88%, and 47% to 61% of that of total Put, Spd, and Spm, respectively. The percentage of bound-to-free PAs in the transgenic tissue had a wide range, varying from 41% to 293%, 5% to 233%, and 56% to 1630% for Put, Spd, and Spm, respectively (Figure 6). The 5 DBP and 2 DAP transgenic tissues contained up to 16-fold higher bound Spm compared to free Spm (Figure 6). These data indicate that expression of *ySpdSyn* under a fruit-specific promoter SlE8 (E8-8 line) or under a constitutive CaMV35S promoter (lines C4, C15) results in large accumulation of bound forms of Put, Spd, and Spm, as compared to the WT.

2.6. Transgene Increased Expression of PA Biosynthesis and Catabolic Pathway Genes

We also quantified the levels of the biosynthesis and catabolic pathway genes of PAs in WT, C4 and E8-8 ovaries at 2 DAP. The steady state transcript levels of *ADC* and *ODC* genes increased significantly in 2 DAP ovaries in both C4 and E8-8 lines (Figure 7a). In 2 DAP ovaries, the transcripts of three tomato *SAMdc* genes were found to be differentially affected, with the steady state levels of *SAMDC2* and *SAMDC3* being higher in E8-8 line and slightly downregulated in the C4 line (Figure 7b). In contrast, the steady state levels of *SAMDC1* transcripts remained unchanged in response to transgene expression. Among the PA catabolic genes, significantly higher levels of *CuAO* and *CuAO-like* gene transcripts accumulated in the C4 line compared to the WT, whereas significantly higher levels of *CuAO-like* and *PAO4-like* gene transcripts were observed in 2 DAP ovaries from E8-8 lines (Figure 7a).

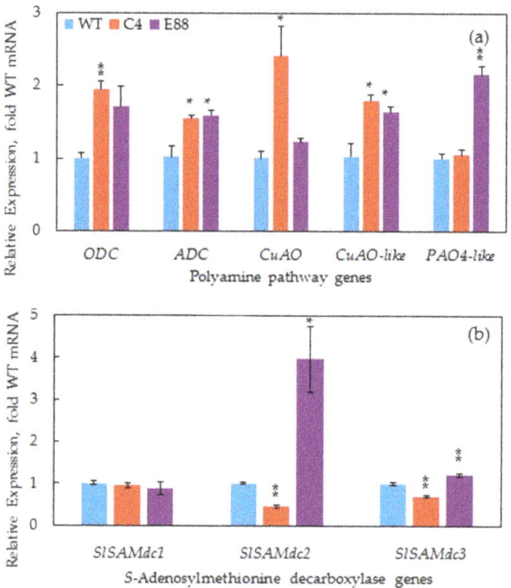

Figure 7. Steady state transcript levels of *ODC*, *ADC*, *Polyamine oxidases* (*CuAo*, *CuAo-like* and *PAO4-like*) and *SAMDC* genes (*SAMDC1*, *SMADC2*, *SAMDC3*) in WT and transgenic tomato ovaries 2 days after pollination. Transcripts were quantified using qRT-PCR and relative expression levels were calculated by the $2^{-\Delta\Delta C_T}$ method using *SlACTIN* (Solyc04g011500.2.1) as housekeeping gene and plotted as fold-respective to WT tissues. Other details were the same as in the Figure 3 legend.

2.7. Statistical Analyses Accentuate Positive Correlations between Specific PAs levels, Gene Transcripts, and Fruit Architectural Parameters During Early Fruit Development

Principal Component Analyses (PCA) of free and bound Put, Spd, and Spm content, together with expression levels of various cell division and cell expansion genes were measured at developmental stages from 2 DAP to 20 DAP (Figure 8). The free and bound forms of all three PAs clustered in the positive half of F1 along with *CYCB2*, *CDKA1*, *CDKB2* (two gene associated with cell division) and their inhibitors *KRP1* and *WEE1* clustered primarily with 2 and 5 DAP, the stages that regulate cell number in developing ovaries (Figure 8a). The observed clustering of cell expansion genes *CCS52A*, *CCS52B*, and *FSM1* predominately with 10 and 20 DAP stages and away from the free and bound forms of three PAs suggests that PA levels are inversely associated with fruit expansion phases (Figure 8a). Among the fruit shape genes, *SUN1* and *OVATE* expression were positively associated with free forms of all the three PAs quantified, whereas *FW2.2* was found to have a negative association (Figure 8a). The number of cell layers and pericarp thickness were associated with 10 and 20 DAP stages of the three transgenic lines, but not with WT fruit (Figure 8b). However, with further progression of ovaries from 10 DAP stage to 20 DAP stage, the number of cell layers and pericarp thickness were associated with both transgenic lines and WT (Figure 8b), suggesting that PAs cause early changes in these parameters.

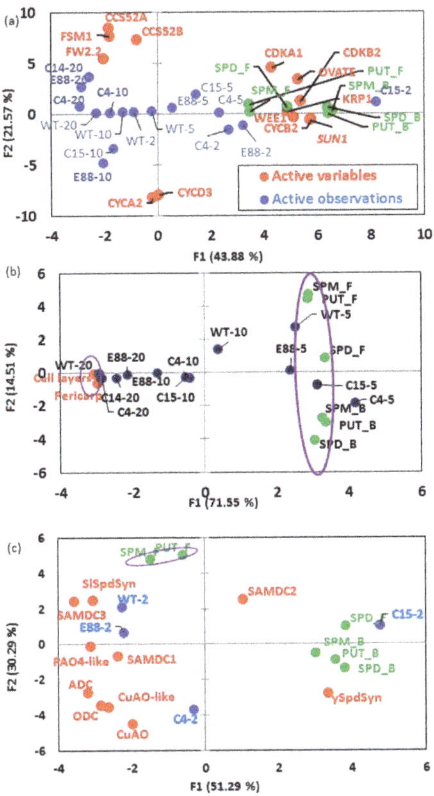

Figure 8. Principal component analyses (PCA) for polyamines (PAs), gene transcripts (cell division, cell expansion, fruit shape-realated and polyamine pathways) and cytological attributes of developing ovaries from WT and transgenic tomato lines. (**a**) PCA between free and bound forms of PAs and transcript levels of genes involved in cell division, cell expansion, and fruit shape development in tomato ovaries at 2, 5, 10 and 20 days after pollination (DAP); (**b**) PCA between free and bound forms of PAs and cell layers, and pericarp thickness of developing tomato ovaries at 5, 10, and 20 DAP; (**c**) PCA between free and bound forms of PAs and transcript levels of genes involved in PA biosynthesis (*ADC*, ODC, *SAMDCs* and *SlSpdSyn/ySpdSyn*) and catabolism (*CuAO*, *CuAO-like*, *PAO4-like*) in tomato ovaries at 2 DAP. Color codes: green—free and bound forms of PAs; blue—genotype and its ovary development stage (within parentheses); red—gene trasncripts or cells cytological attributes. Abbreviations: Put_F/B—free or bound form of Put; Spd_F/B—free or bound form of Spd; Spm_F/B—free or bound Spm.

During the cell division phase (2 DAP), free Spd and bound forms of the three PAs were found positively associated with *ySpdSyn* and *SAMDC2* transcripts and negatively with all other PA biosynthetic as well as catabolic gene transcripts examined (Figure 8c). The Pearson correlations coefficient analysis showed that free Spd levels were positively correlated ($\alpha \geq 0.05$) with transcript levels of *SUN1* and *OVATE*, while bound forms of Put, Spd and Spm were positively correlated with *SUN1*, *OVATE*, *CDKA1*, *CDKB2*, *CYCB2*, *KRP1*, and *WEE1* (Figure S3). Notably, transcripts of fruit shape gene *FW2.2* and cell cycle genes *CCS52A*, *CCS52B*, *FSM1*, *CYCA2*, and *CYCD3* did not have significant correlation with free or bound forms of PAs, suggesting a limited role of these cell cycle genes in affecting phenotype of fruit carrying FW2.2 mutation (Figure S4).

3. Discussion

We show here that PAs are cellular modifiers of fruit shape in tomato. This conclusion is based on the characterization of independent transgenic lines of tomato transformed with a heterologous *ySpdSyn* gene under two different promoters. The shape modification results in a more obovoid fruit shape, as assessed using three independent isogenic transgenic lines in comparison to the relatively spherical round shape of WT fruit (Figure 1a). The structural basis of altered morphology in the transgenic fruit is likely due to the reduction of cellular layers in pericarp, as the thickness of pericarp decreased by 50%, 54%, and 80% of WT in C15, E8-8, and C4 fruits, respectively (Figure 2c,d). Cytological examination of slices from ovaries at different stages of development revealed that PAs are associated with the periclinal cell division, leading to reduced number of cell layers in pericarp and the cell number in medial-lateral direction of pericarp, but not cell size per se. However, we could not conclude that reduced cellular layers and cell size are responsible for the obovoid shape and size of tomato fruit.

More extensive ultrastructure work should enable further understanding of the structural basis of diversity observed in tomato fruit. Altered fruit phenotypes have also been observed in mutants compromised in various growth and development processes, including flowering, fruiting, and photosynthesis, suggesting that many plant genes/proteins/processes influence fruit development [1,39]. Also, breeding for increased fruit size had previously led to large variations in fruit shape of cultivated tomato, from round to oblate, pear-shaped, torpedo-shaped, and bell pepper–shaped [1].

Several genes that affect tomato fruit shape have been identified and characterized [40]. Our results demonstrate that transgenic tomato with enhanced levels of PAs are altered in the expression of *SUN1* and *OVATE* during the early stages of development. Expression of *SUN1* leads to elongated fruit in tomato [17]. Expression of *SUN1* increased in the transgenic fruit, suggesting that it is associated with the elongation of fruit likely causing a more obovoid phenotype (Figure 4). In the mutated form, *OVATE* gene enhances obovoid over the round phenotype in tomato fruit [2]. By cloning *OVATE* mRNA from WT and three transgenic lines and determining DNA sequence of these clones, we have shown that the *OVATE* gene in Ohio 8245 tomato cultivar and its isogenic transgenic lines is a non-mutated type and would, therefore, favor a more spherical round fruit in Ohio 8245 cultivar (Figure S1). Thus, higher expression of this gene cannot be the cause of obovoid phenotype observed in the three transgenic lines. In addition to its phenotype, the *OVATE* gene is known to affect multiple factors in fruit, including metabolome, physiology, and fruit quality [41,42]. *Fw2.2* primarily affects fruit size as studies with transgenic fruit expressing WT *FW2.2* gene showed reduction in the size of fruit cv. Mogeor from a large one to a smaller one [6]. *FW2.2* has been also reported to have pleiotropic effects, especially on the distribution of photosynthate among all fruits on the plant [15]. The levels of *FW2.2* decreased significantly in transgenic fruit from 2 DAP to 20 DAP (Figure 4), which may result in the increase of fruit size and contribute to fruit shape. Expression of *SlSpdSyn* was upregulated during both the cell division and the cell expansion phases of fruit development, suggesting a dual role of Spd in fruit development (Figure 3).

It is generally accepted that the levels of free cellular PAs are under a tight genetic regulation to maintain PAs homeostasis during the growth and development of organisms [26,36,43]. Our data support this hypothesis. However, the role(s) of bound PAs in maintaining PAs homeostasis remains unknown. A nexus between PA levels, ratios of different PAs, and regulation of flower development is known [26]. Several mechanisms have been proposed to maintain the PAs homeostasis in organisms. These include biosynthesis, catabolism, conjugation, and transport of various PAs. Additionally, feedback inhibition and activation of PAs biosynthetic enzymes have been reported and suggested to play a role in establishing PA homeostasis. Due to the cationic nature of various PAs, their binding to various macromolecules in cellular milieu would also affect their cellular homeostasis [24]. Catabolism of PAs not only changes the levels of Put by back conversion of Spd and Spm, but also generates bioactive H_2O_2 [24]. Significant increases in the gene transcripts catalyzing PA catabolism, viz., *CuAO*, *CuAO-like*, and *PAO4-like* amine oxidases, suggest inter-conversion of Spm to Spd and Spd to Put (Figure 6), as reported by others [44,45]. PA depletion has also been implicated in a large number of plant responses and cytotoxic effects of excessive PAs [46,47]. Genetic regulation of PA homeostasis has been suggested as a means to maintain the cellular PAs concentration below toxic levels [43].

Expression of *SlSpdSyn* was upregulated both during the cell division and the cell expansion phase of fruit development, suggesting a dual role of PAs in fruit development (Figure 3). A large increase in bound forms of Put, Spd, and Spm in transgenic ovaries was observed in this study. However, the chemical nature of the bound PAs still needs to be investigated (Figure 6). The bound forms of various PAs could represent formation of conjugates or increased ionic binding to protein, RNA, DNA, and other chemical moieties. The molecular basis of enhanced Put, Spm, and Spm in such bound forms is not known, but would likely result from enhanced biosynthesis of PAs, especially when PA biosynthesis genes are upregulated, as observed in this investigation (Figure 7). However, the role of catabolism in this scenario remains to be determined. Elucidation of the chemical basis of massive accumulation of bound PAs in transgenic fruits should help understand their role in PAs homeostasis. In this context, it is known that PAs induce Spd/Spm *N1*-acetyltransferase and efflux transporters and likely self-regulate their levels [48–51].

4. Materials and Methods

4.1. Plant Material and Growth Conditions

Wild type and transgenic tomato lines homozygous for *ySpdSyn* gene fused to either constitutive CaMV35S promoter (lines C4 and C15) or fruit/ethylene-specific SlE8 promoter (line E8-8) have been described previously [27]. These plants were grown in high porosity potting mix (52Mix, Conrad Fafard, Inc., Agawam, MA, USA) in a greenhouse with 16 h day/8 h night photoperiod and 23 °C day/18 °C night temperature. Tomato flowers and fruit developmental stages were tagged. Samples were collected on day 10 and day 5 before pollination (DBP) and 2, 5, 10, and 20 days after pollination (DAP) (Figure S4), immediately frozen in liquid N_2 and stored at −80 °C until further use. To determine fruit shape index, fruits at various stages of development were sliced from proximal-distal axis, scanned, and analyzed using 'Tomato Analyzer' software [52].

4.2. Cytological Analysis

Axis slices of fresh tissues were fixed in 10% formalin (pH 6.8–7.2) and processed in Tissue-Tek VIP® (Sakura Finetek USA Inc., Torrance, CA, USA) using following sequential treatments: 70% ethanol for 50 s, 95% ethanol for 50 s (2 cycles), 100% ethanol for 33 s (3 cycles), toluene for 60 s (2 cycles), and paraffin for 45 s at 63 °C (4 cycles). For deparaffinization, tissues were dipped in xylene for 5 min (2 cycles), 100% ethanol for 2 min, 95% ethanol for 2 min, and 70% ethanol for 2 min, followed by rehydration in deionized water. Tissues were embedded in paraffin using Cryo-therm (Lipshaw, Tucker, GA, USA) and 5–7 μm thick sections were made using microtome (Finesse ME, Thermo Electron, Waltham, MA, USA). These were then stained with 1% toluidine blue–O, sectioned, dehydrated by serial quick dips in 70%, 95%, and 100% ethanol, and xylene [53]. Slides were scanned with ScanScope CS (Aperio Technologies Inc., Buffalo Grove, IL, USA) at different magnifications, 40x being the maximum. The digital images were analyzed for cell number and cell layers using

ImageScope 11 (Aperio Technologies Inc., Buffalo Grove, IL, USA), and for cell size by ImageJ [54]. Three independent biological replicates were analyzed at each stage from each genotype.

4.3. Transcript Analysis by Quantitative Real-Time PCR

For qRT-PCR, the frozen tissues were ground to powder using liquid N_2, total RNA was extracted from 100 mg tissue powder using QIAzol® Lysis reagent (Qiagen Sciences, Germantown, MD, USA) and purified using RNeasy® Mini Kit (Qiagen Sciences, Germantown, MD, USA). The RNA samples were treated with one unit RQ1 RNase-free DNase (Promega Corporation, Madison, WI, USA) and first-strand cDNA was synthesized using SuperScript II Reverse Transcriptase (Invitrogen, Waltham, MA, USA). GoTaq® qPCR Master Mix (Promega Corporation, Madison, WI, USA) was used in qRT-PCR reaction mixture. All primers and cDNA templates were optimized for gene expression analysis according to the $2^{-\Delta\Delta CT}$ method [55]. StepOnePlus™ Real-Time PCR System (Applied Biosystems, Waltham, MA, USA) was used with program sequence as follows: 95°C for 10 min; 95 °C for 15 s and 60 °C for 60 sec (40 cycles); 95 °C for 15 s; and 60 °C for 60 s. Comparative C_T values of gene expression were quantified using StepOne™ 2.0 software (Applied Biosystems, Waltham, MA, USA). *ACTIN* was used as standard housekeeping gene to normalize the expression of target genes. Accession numbers of all genes with their primer sequences used in this study are listed in Table S3. All data presented here represent average ± standard error of a minimum of three independent biological replicates.

4.4. Quantification of Polyamines by High Pressure Liquid Chromatography

Polyamines were extracted from floral buds and fruit tissues of tomato plants and dansylated as described previously [56] with some modifications. Briefly, 200 mg of finely ground sample was homogenized in 800 µL of 5% ice-cold perchloric acid (PCA) using a hand-held homogenizer precooled at 4 °C for 1 h. The homogenate was centrifuged at 20,000 *g* for 30 min at 4 °C and both the supernatant and residue were collected separately. PAs present in the supernatant were labeled as free PAs. To quantify PCA-insoluble bound PAs, the 20,000 *g* pellet was washed twice with 5% PCA, re-suspended in 800 µL of 5% cold PCA, and then hydrolyzed in an equal volume of 6N HCl for 18 h at 110 °C. Saturated sodium carbonate (200 µL) and 1,7-heptanediamine (400 µL, as an internal standard) were added to the 100 µL supernatant or the hydrolyzates, and then dansylated with dansyl chloride for 60 min at 60 °C in the dark. Dansylation was terminated by adding 100 µL proline and incubating the reaction mixture for 30 min at 60 °C. Dansylated PAs were extracted in 500 µL toluene, air dried, and dissolved in 250 µL acetonitrile. The samples were then diluted four times with acetonitrile, filtered through a 0.45 µm syringe filter (National Scientific, Claremont, CA, USA), and fractionated on a reversed-phase Nova-Pak C18 column (3.9 × 150 mm, 4.0 µm pore size) using Waters 2695 Separation Module equipped with Waters 2475 Multi λ fluorescence detector (excitation 340 nm, emission 510 nm) with a binary gradient composed of solvent A (100% Water) and solvent B (100% acetonitrile) at 1 mL/min. Initial conditions were set at 60:40 (A:B) and then a linear gradient was processed with conditions set at 30:70 (A:B) at 3 min; 0:100 (A:B) at 10 min, and 60:40 (A:B) at 12 min. The column was flushed with 60:40 (A:B) for 3 min before the next sample injection. To determine PAs recovery and generate calibration curves, authentic PA standards (Sigma-Aldrich, St. Louis, MO, USA) were used as control. PAs were integrated and quantified using Millennium³² 4.0 from Waters Corporation. After hydrolysis, PCA-soluble and PCA-insoluble samples were quantified and are designated as free and bound forms of each PA, respectively, throughout the manuscript.

5. Conclusions

Our findings on gene transcripts regulating cell division and cell enlargement in response to the expression of *ySpdSyn* transgene and subsequent alteration in PAs levels are summarized in a model (Figure 9). The cellular levels of bound Put, Spd, and Spm correlate with *CDKA1, CDKB2, CYCB2, KRP1,* and *WEE1* transcripts, but not *CCS52A, CCS52B, CYCA2,* and *CYCD3* genes in transgene-associated change in fruit shape. The patterns of gene expression during early fruit development indicate that the upregulated *CDKB2* may in turn activate *CYCB2* expression during ovary development at 2 DAP and 5 DAP stages, the phase of active cell division of pollinated ovaries. The precipitous decrease in the expression of all cyclins in 20 DAP tomato fruit is consistent with the end of cell division phase at this stage of fruit development [37]. It is noted here that fruit shape changes occur at an early stage of fruit development (before 5 DAP), during enhanced accumulation of PAs, and continue to persist during fruit development and ripening, as was also found to be the case for *SUN1* and *OVATE* transgenic fruits. Our results indicate that the expression of the transgene is associated with reduced cell division in the medial-lateral direction of fruit, which can cause more elongated fruit shape, an effect also seen in tomato fruit over-expresser for *SUN1* gene [17,57]. In our studies, the parental Ohio 8245 tomato cultivar expressed a functional *OVATE* gene, which would reduce cell division at the distal end and induce more round fruit shape. Thus, increase in obovoid shape of the transgenic fruit seems to be linked to both *SUN1* and *OVATE* genes, with *SUN1* likely having a stronger effect on the fruit phenotype.

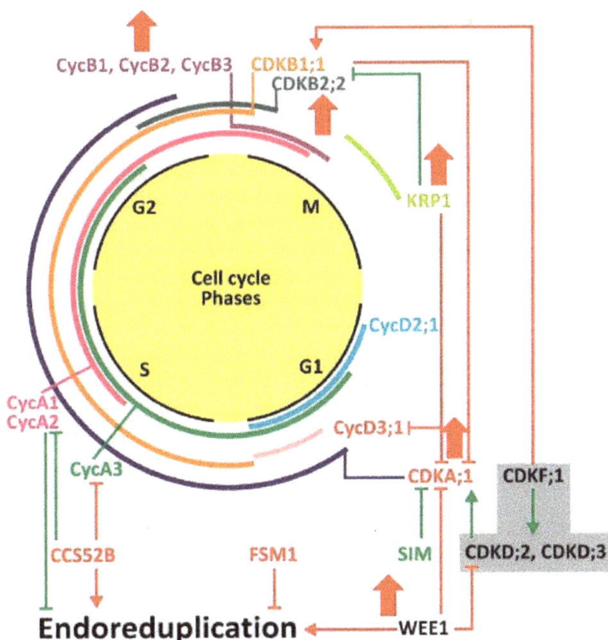

Figure 9. A model showing the polyamine-mediated changes in transcript levels of genes involved in cell cycle progression and endoreduplication during tomato fruit development. In eukaryotes, cell cycle is mainly comprised of interphase and mitosis (M phase). Interphase is further divided into three phases: i) G1 (Gap 1) during which cell increases in size and becomes ready for DNA synthesis; ii) S (Synthesis) where DNA replication occurs; and iii) G2 (Gap 2) during which cell either continues to grow until ready for mitosis or enter into DNA amplification phase, called endoreduplication [58,59]. Expression levels of cyclins or CDKs during different phases of cell cycle progress are indicated in colors. CDK-activating kinases are highlighted in gray box. Transcript levels of genes enhanced by higher polyamines (Spd/Spm) are indicated with upward-facing thick red arrows. Abbreviations: G—Gap phase; M—Mitosis phase; S—Synthesis phase; Cyc—Cyclins; CDKs—Cyclin dependent kinases; KRPs—Kip-related proteins.

Supplementary Materials: The following are available online at http://www.mdpi.com/2223-7747/8/10/387/s1, Figure S1: Sequence alignment of *OVATE* gene in WT and transgenic fruits with its mutated version (*ovate*) containing stop codon (TAA); Figure S2: Expression patterns of various cell division and cell expansion genes in WT developing floral and ovary tomato tissues; Figure S3: Correlation coefficient analysis of PA levels and transcripts of genes involved in fruit shape, cell cycle progression, cell expansion and polyamine biosynthesis and catabolism; Figure S4: Representative flower and fruit developmental stages registered in *Solanum lycopersicum* cv. Ohio8245; Table S1: Steady state transcript levels of cell division, cell expansion, fruit shape and polyamine pathway genes at various developmental stages of floral and ovaries tissues in transgenic tomato line C15; Table S2: Free and bound forms of putrescine, spermidine and spermine in floral buds and ovaries tissues at various developmental from wild-type (WT) and *35S:ySpdSyn* expressing transgenic tomato line C15; Table S3: List of genes and their primer sequences used for quantitative real-time PCR analyses. References [60] are cited in the Supplementary Materials.

Author Contributions: Conceptualization, A.K.H. and R.A.; Data curation, R.A. and S.F.; Formal analysis, A.K.M. and A.K.H.; Funding acquisition, A.K.M. and A.K.H.; Investigation, R.A.; Methodology, S.F.; Project administration, A.K.H.; Supervision, A.K.H.; Writing—original draft, R.A. and A.K.H.; Writing—review & editing, R.A., A.K.M. and A.K.H.

Funding: The studies were partly supported by a US–Israel Binational Agricultural Research and Development Fund to AKH and AKM (Grant No. IS-3441-03), and USDA-IFAFS program grant (Award No. 741740) to AKH. RA was partially supported by the Higher Education Commission, Islamabad (Pakistan). AKM is supported through USDA-ARS intramural Project No: 8042-21000-143-00D. The mention of trade names or commercial products in this publication is solely for the purpose of providing specific information and does not imply recommendation or endorsement by the US Department of Agriculture.

Conflicts of Interest: The authors declare no conflict of interest.

References

1. Tanksley, S.D. The genetic, developmental, and molecular bases of fruit size and shape variation in tomato. *Plant Cell* **2004**, *16*, S181–S189. [CrossRef] [PubMed]

2. Liu, J.; Van Eck, J.; Cong, B.; Tanksley, S.D. A new class of regulatory genes underlying the cause of pear-shaped tomato fruit. *Proc. Natl. Acad. Sci. USA* **2002**, *99*, 13302–13306. [CrossRef]

3. Brewer, M.T.; Moyseenko, J.B.; Monforte, A.J.; van der Knaap, E. Morphological variation in tomato: A comprehensive study of quantitative trait loci controlling fruit shape and development. *J. Exp. Bot.* **2007**, *58*, 1339–1349. [CrossRef] [PubMed]

4. Handa, A.K.; Anwar, R.; Mattoo, A.K. Biotechnology of fruit quality. In *Fruit Ripening: Physiology, Signalling and Genomics*; Nath, P., Bouzayen, M., Mattoo, A.K., Pech, J.C., Eds.; CAB International: Oxfordshire, UK, 2014; pp. 259–290.

5. Lin, T.; Zhu, G.; Zhang, J.; Xu, X.; Yu, Q.; Zheng, Z.; Zhang, Z.; Lun, Y.; Li, S.; Wang, X.; et al. Genomic analyses provide insights into the history of tomato breeding. *Nat. Genet.* **2014**, *46*, 1220. [CrossRef]

6. Frary, A.; Nesbitt, T.C.; Frary, A.; Grandillo, S.; Van Der Knaap, E.; Cong, B.; Liu, J.; Meller, J.; Elber, R.; Alpert, K.B.; et al. Fw2.2: A quantitative trait locus key to the evolution of tomato fruit size. *Science* **2000**, *289*, 85–88. [CrossRef] [PubMed]

7. Cong, B.; Barrero, L.S.; Tanksley, S.D. Regulatory change in yabby-like transcription factor led to evolution of extreme fruit size during tomato domestication. *Nat. Genet.* **2008**, *40*, 800–804. [CrossRef] [PubMed]

8. Xiao, H.; Jiang, N.; Schaffner, E.; Stockinger, E.J.; van der Knaap, E. A retrotransposon-mediated gene duplication underlies morphological variation of tomato fruit. *Science* **2008**, *319*, 1527–1530. [CrossRef]

9. Huang, Z.J.; van der Knaap, E. Tomato fruit weight 11.3 maps close to fasciated on the bottom of chromosome 11. *Theor. Appl. Genet.* **2011**, *123*, 465–474. [CrossRef]

10. Muños, S.; Ranc, N.; Botton, E.; Bérard, A.; Rolland, S.; Duffé, P.; Carretero, Y.; Le Paslier, M.-C.; Delalande, C.; Bouzayen, M.; et al. Increase in tomato locule number is controlled by two single-nucleotide polymorphisms located near wuschel. *Plant Physiol.* **2011**, *156*, 2244–2254. [CrossRef]

11. Rodríguez, G.R.; Muños, S.; Anderson, C.; Sim, S.-C.; Michel, A.; Causse, M.; Gardener, B.B.M.; Francis, D.; van der Knaap, E. Distribution of sun, ovate, lc, and fas in the tomato germplasm and the relationship to fruit shape diversity. *Plant Physiol.* **2011**, *156*, 275–285. [CrossRef]

12. Clevenger, J. Metabolic and Genomic Analysis of Elongated Fruit Shape in Tomato (*Solanum Lycopersicum*). Master's Thesis, The Ohio State University, Horticulture and Crop Science, Columbus, OH, USA, 2012.

13. Chakrabarti, M.; Zhang, N.; Sauvage, C.; Muños, S.; Blanca, J.; Cañizares, J.; Diez, M.J.; Schneider, R.; Mazourek, M.; McClead, J.; et al. A cytochrome p450 regulates a domestication trait in cultivated tomato. *Proc. Natl. Acad. Sci. USA* **2013**, *110*, 17125–17130. [CrossRef] [PubMed]

14. Alpert, K.B.; Grandillo, S.; Tanksley, S.D. *Fw 2.2*: A major qtl controlling fruit weight is common to both red- and green-fruited tomato species. *TAG Theor. Appl. Genet.* **1995**, *91*, 994–1000. [CrossRef] [PubMed]

15. Nesbitt, T.C.; Tanksley, S.D. Comparative sequencing in the genus lycopersicon: Implications for the evolution of fruit size in the domestication of cultivated tomatoes. *Genetics* **2002**, *162*, 365–379. [PubMed]

16. van der Knaap, E.; Tanksley, S.D. Identification and characterization of a novel locus controlling early fruit development in tomato. *Theor. Appl. Genet.* **2001**, *103*, 353–358. [CrossRef]

17. Wu, S.; Xiao, H.; Cabrera, A.; Meulia, T.; van der Knaap, E. *Sun* regulates vegetative and reproductive organ shape by changing cell division patterns. *Plant Physiol.* **2011**, *157*, 1175–1186. [CrossRef]

18. Wang, Y.; Clevenger, J.P.; Illa-Berenguer, E.; Meulia, T.; van der Knaap, E.; Sun, L. A comparison of *sun, ovate, fs8.1* and auxin application on tomato fruit shape and gene expression. *Plant Cell Physiol.* **2019**, *60*, 1067–1081. [CrossRef]

19. Ku, H.M.; Doganlar, S.; Chen, K.Y.; Tanksley, S.D. The genetic basis of pear-shaped tomato fruit. *TAG Theor. Appl. Genet.* **1999**, *99*, 844–850. [CrossRef]

20. Tsaballa, A.; Pasentsis, K.; Darzentas, N.; Tsaftaris, A. Multiple evidence for the role of an ovate-like gene in determining fruit shape in pepper. *BMC Plant Biol.* **2011**, *11*, 46. [CrossRef]

21. Tiburcio, A.F.; Alcazar, R. Potential Applications of Polyamines in Agriculture and Plant Biotechnology. *Methods Mol. Biol.* **2018**, *1694*, 489–508.

22. Pegg, A.E. Introduction to the Thematic Minireview Series: Sixty plus years of polyamine research. *J. Biol. Chem.* **2018**, *293*, 18681–18692. [CrossRef]

23. Miyamoto, M.; Shimao, S.; Tong, W.; Motose, H.; Takahashi, T. Effect of Thermospermine on the Growth and Expression of Polyamine-Related Genes in Rice Seedlings. *Plants* **2019**, *8*, 269. [CrossRef] [PubMed]

24. Handa, A.K.; Fatima, T.; Mattoo, A.K. Polyamines: Bio-molecules with diverse functions in plant and human health and disease. *Front. Chem.* **2018**, *6*, 10. [CrossRef] [PubMed]

25. Seifi, H.S.; Shelp, B.J. Spermine differentially refines plant defense responses against biotic and abiotic stresses. *Front. Plant Sci.* **2019**, *10*, 117. [CrossRef] [PubMed]

26. Handa, A.K.; Nambeesan, S.; Mattoo, A.K. Nexus between spermidine and floral organ identity and fruit/seed set in tomato. *Front. Plant Sci.* **2019**, *10*, 1033.

27. Nambeesan, S.; Datsenka, T.; Ferruzzi, M.G.; Malladi, A.; Mattoo, A.K.; Handa, A.K. Overexpression of yeast spermidine synthase impacts ripening, senescence and decay symptoms in tomato. *Plant J.* **2010**, *63*, 836–847. [CrossRef] [PubMed]

28. Mattoo, A.K.; Sobieszczuk-Nowicka, E. Polyamine as signaling molecules and leaf senescence. In *Senescence Signalling and Control in Plants*; Sarwat, M., Tuteja, N., Eds.; Academic Press: Cambridge, MA, USA, 2019; pp. 125–138.

29. Chen, D.; Shao, Q.; Yin, L.; Younis, A.; Zheng, B. Polyamine function in plants: Metabolism, regulation on development, and roles in abiotic stress responses. *Front. Plant Sci.* **2019**, *9*, 1945. [CrossRef] [PubMed]

30. Ikeuchi, M.; Ogawa, Y.; Iwase, A.; Sugimoto, K. Plant regeneration: Cellular origins and molecular mechanisms. *Development* **2016**, *143*, 1442–1451. [CrossRef]

31. Mehta, R.A.; Cassol, T.; Li, N.; Ali, N.; Handa, A.K.; Mattoo, A.K. Engineered polyamine accumulation in tomato enhances phytonutrient content, juice quality, and vine life. *Nat. Biotechnol.* **2002**, *20*, 613–618. [CrossRef] [PubMed]

32. Jang, S.; Cho, H.; Park, K.; Kim, Y.-B. Changes in cellular polyamine contents and activities of their biosynthetic enzymes at each phase of the cell cycle in by-2 cells. *J. Plant Biol* **2006**, *49*, 153–159. [CrossRef]

33. Verkest, A.; Weinl, C.; Inzé, D.; De Veylder, L.; Schnittger, A. Switching the cell cycle. Kip-related proteins in plant cell cycle control. *Plant Physiol.* **2005**, *139*, 1099–1106. [CrossRef]

34. Martin-Tanguy, J. Metabolism and function of polyamines in plants: Recent development (new approaches). *Plant Growth Regul.* **2001**, *34*, 135–148. [CrossRef]

35. Tang, W.; Newton, R. Polyamines promote root elongation and growth by increasing root cell division in regenerated virginia pine (*pinus virginiana* mill.) plantlets. *Plant Cell Rep.* **2005**, *24*, 581–589. [CrossRef] [PubMed]

36. Anwar, R.; Mattoo, A.K.; Handa, A.K. Polyamine interactions with plant hormones: Crosstalk at several levels. In *Polyamines: A Universal Molecular Nexus for Growth, Survival, and Specialized Metabolism*; Kusano, T., Suzuki, H., Eds.; Springer: London, UK, 2015; pp. 267–302.

37. Czerednik, A.; Busscher, M.; Angenent, G.C.; de Maagd, R.A. The cell size distribution of tomato fruit can be changed by overexpression of cdka1. *Plant Biotechnol. J.* **2015**, *13*, 259–268. [CrossRef] [PubMed]

38. Barg, R.; Sobolev, I.; Eilon, T.; Gur, A.; Chmelnitsky, I.; Shabtai, S.; Grotewold, E.; Salts, Y. The tomato early fruit specific gene *lefsm1* defines a novel class of plant-specific sant/myb domain proteins. *Planta* **2005**, *221*, 197–211. [CrossRef] [PubMed]

39. Balbi, V.; Lomax, T.L. Regulation of early tomato fruit development by the diageotropica gene. *Plant Physiol.* **2003**, *131*, 186–197. [CrossRef]

40. Azzi, L.; Deluche, C.; Gévaudant, F.; Frangne, N.; Delmas, F.; Hernould, M.; Chevalier, C. Fruit growth-related genes in tomato. *J. Exp. Bot.* **2015**, *66*, 1075–1086. [CrossRef]

41. Wang, S.; Chang, Y.; Guo, J.; Zeng, Q.; Ellis, B.E.; Chen, J.-G. Arabidopsis ovate family proteins, a novel transcriptional repressor family, control multiple aspects of plant growth and development. *PLoS ONE* **2011**, *6*, e23896. [CrossRef]

42. Liu, J.; Zhang, J.; Wang, J.; Zhang, J.; Miao, H.; Jia, C.; Wang, Z.; Xu, B.; Jin, Z. *Mumads1* and *maofp1* regulate fruit quality in a tomato ovate mutant. *Plant Biotechnol. J.* **2018**, *16*, 989–1001. [CrossRef]

43. Fellenberg, C.; Handrick, V.; Ziegler, J.; Vogt, T. Polyamine homeostasis in wild type and phenolamide deficient *arabidopsis thaliana* stamens. *Front. Plant Sci.* **2012**, *3*, 180. [CrossRef]

44. Moschou, P.N.; Paschalidis, K.A.; Roubelakis-Angelakis, K.A. Plant polyamine catabolism: The state of the art. *Plant Signal. Behav.* **2008**, *3*, 1061–1066. [CrossRef]

45. Takahashi, T.; Tong, W. Regulation and diversity of polyamine biosynthesis in plants. In *Polyamines: A Universal Molecular Nexus for Growth, Survival, and Specialized Metabolism*; Kusano, T., Suzuki, H., Eds.; Springer: Japan, Tokyo, 2015; pp. 27–44.

46. Tobias, K.E.; Kahana, C. Exposure to ornithine results in excessive accumulation of putrescine and apoptotic cell-death in ornithine decarboxylase overproducing mouse myeloma cells. *Cell Growth Differ.* **1995**, *6*, 1279–1285. [PubMed]

47. Handa, A.K.; Mattoo, A.K. Differential and functional interactions emphasize the multiple roles of polyamines in plants. *Plant Physiol. Biochem.* **2010**, *48*, 540–546. [CrossRef] [PubMed]

48. Sandusky-Beltran, L.A.; Kovalenko, A.; Ma, C.; Calahatian, J.I.T.; Placides, D.S.; Watler, M.D.; Hunt, J.B.; Darling, A.L.; Baker, J.D.; Blair, L.J.; et al. Spermidine/spermine-n1-acetyltransferase ablation impacts tauopathy-induced polyamine stress response. *Alzheimer's Res. Ther.* **2019**, *11*, 58. [CrossRef] [PubMed]

49. Casero, R.A.; Pegg, A.E. Spermidine/spermine n1-acetyltransferase–the turning point in polyamine metabolism. *FASEB J.* **1993**, *7*, 653–661. [CrossRef] [PubMed]

50. Sobieszczuk-Nowicka, E.; Paluch-Lubawa, E.; Mattoo, A.K.; Arasimowicz-Jelonek, M.; Gregersen, P.L.; Pacak, A. Polyamines—A new metabolic switch: Crosstalk with networks involving senescence, crop improvement, and mammalian cancer therapy. *Front. Plant Sci.* **2019**, *10*, 859. [CrossRef] [PubMed]

51. Pooja; Goyal, V.; Devi, S.; Munjal, R. Role of polyamines in protecting plants from oxidative stress. In *Reactive Oxygen, Nitrogen and Sulfur Species in Plants: Production, Metabolism, Signaling and Defense Mechanisms*; Hasanuzzaman, M., Fotopoulos, V., Nahar, K., Fujita, M., Eds.; John Wiley & Sons: Hoboken, NJ, USA, 2019; pp. 143–157.

52. Gonzalo, M.J.; Brewer, M.T.; Anderson, C.; Sullivan, D.; Gray, S.; van der Knaap, E. Tomato fruit shape analysis using morphometric and morphology attributes implemented in tomato analyzer software program. *J. Am. Soc. Hortic. Sci.* **2009**, *134*, 77–87. [CrossRef]

53. Sheehan, D.C.; Hrapchak, B.B. *Theory and Practice of Histotechnology*; Battelle Press: Columbus, OH, USA, 1987; Volume II, p. 481.

54. Schneider, C.A.; Rasband, W.S.; Eliceiri, K.W. Nih image to imagej: 25 years of image analysis. *Nat. Methods* **2012**, *9*, 671–675. [CrossRef]

55. Livak, K.J.; Schmittgen, T.D. Analysis of Relative Gene Expression Data Using Real-Time Quantitative PCR and the $2^{-\Delta\Delta CT}$ Method. *Methods* **2001**, *25*, 402–408. [CrossRef] [PubMed]

56. Torrigiani, P.; Bressanin, D.; Beatriz Ruiz, K.; Tadiello, A.; Trainotti, L.; Bonghi, C.; Ziosi, V.; Costa, G. Spermidine application to young developing peach fruits leads to a slowing down of ripening by impairing ripening-related ethylene and auxin metabolism and signaling. *Physiol. Plant* **2012**, *146*, 86–98. [CrossRef]

57. Xiao, L.; Cui, Y.-H.; Rao, J.N.; Zou, T.; Liu, L.; Smith, A.; Turner, D.J.; Gorospe, M.; Wang, J.-Y. Regulation of cyclin-dependent kinase 4 translation through cug-binding protein 1 and microrna-222 by polyamines. *Mol. Biol. Cell* **2011**, *22*, 3055–3069. [CrossRef]

58. Gutierrez, C. The Arabidopsis cell division cycle. *Arab. Book* **2009**, *7*, e0120. [CrossRef] [PubMed]
59. John, P.C.L.; Qi, R. Cell division and endoreduplication: Doubtful engines of vegetative growth. *Trends Plant Sci.* **2008**, *13*, 121–127. [CrossRef] [PubMed]
60. Mathieu-Rivet, E.; Gévaudant, F.; Sicard, A.; Salar, S.; Do, P.T.; Mouras, A.; Fernie, A.R.; Gibon, Y.; Rothan, C.; Chevalier, C.; et al. Functional analysis of the anaphase promoting complex activator ccs52a highlights the crucial role of endo-reduplication for fruit growth in tomato. *Plant J.* **2010**, *62*, 727–741. [CrossRef] [PubMed]

Article

Variations of Secondary Metabolites among Natural Populations of Sub-Antarctic *Ranunculus* Species Suggest Functional Redundancy and Versatility

Bastien Labarrere [1], Andreas Prinzing [1], Thomas Dorey [2], Emeline Chesneau [1] and Françoise Hennion [1,*]

[1]   UMR 6553 ECOBIO, Université de Rennes 1, OSUR, CNRS, Av du Général Leclerc, F-35042 Rennes, France
[2]   Institut für Systematische und Evolutionäre Botanik, Zollikerstrasse 107, 8008 Zürich, Switzerland
*   Correspondence: francoise.hennion@univ-rennes1.fr

Received: 24 May 2019; Accepted: 16 July 2019; Published: 19 July 2019

Abstract: Plants produce a high diversity of metabolites which help them sustain environmental stresses and are involved in local adaptation. However, shaped by both the genome and the environment, the patterns of variation of the metabolome in nature are difficult to decipher. Few studies have explored the relative parts of geographical region versus environment or phenotype in metabolomic variability within species and none have discussed a possible effect of the region on the correlations between metabolites and environments or phenotypes. In three sub-Antarctic *Ranunculus* species, we examined the role of region in metabolite differences and in the relationship between individual compounds and environmental conditions or phenotypic traits. Populations of three *Ranunculus* species were sampled across similar environmental gradients in two distinct geographical regions in îles Kerguelen. Two metabolite classes were studied, amines (quantified by high-performance liquid chromatography and fluorescence spectrophotometry) and flavonols (quantified by ultra-high-performance liquid chromatography with triple quadrupole mass spectrometry). Depending on regions, the same environment or the same trait may be related to different metabolites, suggesting metabolite redundancy within species. In several cases, a given metabolite showed different or even opposite relations with the same environmental condition or the same trait across the two regions, suggesting metabolite versatility within species. Our results suggest that metabolites may be functionally redundant and versatile within species, both in their response to environments and in their relation with the phenotype. These findings open new perspectives for understanding evolutionary responses of plants to environmental changes.

Keywords: *Ranunculus biternatus*; *Ranunculus pseudotrullifolius*; *Ranunculus moseleyi*; secondary metabolite variation; amines; quercetins; natural populations; environment; redundancy; sub-Antarctic plants

---

1. Introduction

Plants are sessile organisms that have to face changes in their environment. The metabolome stands at an interface between perception of environmental signals and their translation in life history traits, therefore playing a major role in allowing the organism to sustain environmental constraints. Plant metabolites were selected through long-term adaptation and diversification [1]. While primary metabolites (e.g., protein amino acids, carbohydrates), defined as essential to plant physiology are relatively few, the secondary metabolites are especially diverse [2]. Secondary metabolites play major roles in the interactions between the plant and its environment [3], being involved in protection against environmental stresses, competition, or pollinator attraction [2,4,5] and some in vegetative or floral development [6,7]. While the macroevolution of secondary metabolites is becoming more and more precisely deciphered [1,8–12], in contrast relatively little is known about their microevolution [13,14]. In particular, the patterns of variation of the metabolome in nature need to be further investigated.

In nature, plants form populations and in general, plants within populations are more closely related while plants among populations may show stronger evolutionary differences. Equally, populations within regions often tend to be more closely related to each other than populations among regions. Variability in the metabolite composition among populations within species is known for many plant species and metabolites [15–22]. Larger-scale variation such as among regions or latitudes was also investigated, sometimes with the aim to enrich biochemical resources. Thus, studies also show variability in plant metabolite composition across latitudinal gradients or regions [23,24]. Yet, environments may be more different among populations or regions as within, and differences in metabolite composition among populations or regions might hence be dependent or independent of environments. However, relatively few studies have examined differences in metabolite composition among populations or regions, trying to dissociate metabolite variations that are dependent on environments and such that are independent [23–27]. This latter part of variation in metabolite composition that would be independent of environmental factors would likely reflect either selection by unknown past environments or neutral microevolutionary differentiation among populations or regions [13,14] and remains largely to be described [9,28].

High diversity of secondary metabolites has already been reported to correlate with functional diversity [2,5,29]. First, functional convergence is observed through metabolite macroevolution [1,7]. Depending on lineage or species, different metabolites may respond to the same environment or be related to the same trait, called "functional redundancy". Such functional redundancy is observed across all metabolites and plant lineages [1]. Yet, functional redundancy of secondary metabolites might also be observed among plants within species, but this remains to be investigated [30]. Secondly, metabolite versatility has also been observed: a given metabolite may have different roles within a given plant species in different organs or in different environments [1,31,32]. However, metabolite versatility has had little attention. Notably, whether redundancy or versatility might also emerge among populations that occupy similar environments but distant regions remains unknown. We hypothesize that, within species, neutral microevolutionary differentiation among distant but environmentally similar regions exists, and it includes the function of metabolites. Specifically, we hypothesize that functional redundancy or versatility exist within species among plant populations occupying such distant regions. A first step to test these hypotheses would be to search for patterns of correlations in nature that are consistent with functional redundancy or versatility of metabolites, paving the ground for later investigation of their physiological functions. For such a correlational approach, we predict that depending on the region, different metabolites might be correlated to the same environmental factor or the same trait in plants, suggesting functional redundancy. Moreover, depending on the region, the same metabolite might be correlated to different environmental factors or different traits in plants, suggesting functional versatility.

We aim to determine differences in secondary metabolite composition or function in natural populations of plants located in distant but environmentally similar regions. This aim requires several populations of given species distributed across wide environmental gradients within each of multiple regions. These conditions are present in sub-Antarctic Iles Kerguelen [26]. Located in the southern Indian Ocean the Iles Kerguelen harbor a wide range of abiotic environmental conditions [33] and distinct regions across which plants are distributed. This also requires metabolites that are known to vary and play roles in the response to environment or in traits. Among other metabolites, such is the case with amines and flavonoids [34,35]. Previous work comparing the amine metabolomes of nine species showed that amine composition of populations of plants in Iles Kerguelen varied in relation to both species and the environment [26]. Furthermore, previous work on flavonoids in the three *Ranunculus* species growing in Iles Kerguelen showed that quercetins were the only flavonols in these species and that composition of populations varied in relation to both species and the environment [36]. Amines include aliphatic polyamines, acetyl conjugates, and aromatic amines. Polyamines are low molecular weight polycationic molecules with amino groups [7,37] and are described as "growth regulators", being involved in various internal processes such as growth control, DNA replication and cell differentiation, and organ development [3,34]. Polyamines, and aromatic amines by now, are also known to respond to external environment and be involved in the protection of plants against abiotic stresses [7,37,38]. Quercetins are compounds with variable phenolic structures; they belong to flavonols among flavonoids [35]. Quercetins are mainly antioxidants and are involved in the protection of plants against abiotic stresses [35]. A trade-off between flavonoid (including quercetin) concentration and growth is commonly suggested [39]. Therefore, to address the complex variation of plant metabolite composition across populations, environments, traits, and regions in nature, we performed targeted analyses of amines and quercetins. Using two independent metabolite families, we aimed at reinforcing our conclusions.

We sampled populations of the three *Ranunculus* species native to Iles Kerguelen across environmental gradients in two different geographical regions. We focused on environmental factors known to have major effects on these plants, i.e., soil hydric conditions [36,40]. We measured plant metabolite composition (amines and quercetins) and morphological phenotype, and characterized the environment (soil hydric conditions) of populations. We asked whether (i) environment alone explains differences in metabolite composition among populations within species or whether environment-independent microevolutionary differentiation among regions also shapes metabolite composition, (ii) metabolites show patterns consistent with redundancy within species, different metabolites being related to the same environments or morphological traits in the two different regions, (iii) metabolites show patterns consistent with versatility within species, the same metabolites being related to different environments or morphological traits in the two different regions.

2. Results

2.1. Populations Differ in Total Metabolite Contents and Metabolite Composition Partly Independently of the Environment

To identify an effect of population on total metabolite contents that was not due to environmental differences among populations, we conducted an ANCOVA with environmental conditions as co-variables. We found that total contents of amines or quercetins differed significantly among populations across regions and within regions in most cases, i.e., statistically independent of the environmental variables considered (Table 1 and Figure 1).

Table 1. Differences in the total contents of amines or quercetins among populations *.

Amines		Population		
		p-Value	*F* Value	df Residuals
R. biternatus	across regions	**0.001**	4.794	44
	Isthme Bas	**0.004**	10.773	18
	Australia	**0.010**	6.565	24
R. pseudotrullifolius	across regions	**<0.001**	8.722	48
	Isthme Bas	>0.500	0.073	13
	Australia	0.005	6.716	24
R. moseleyi	across regions	**0.001**	4.962	37
	Isthme Bas	0.046	6.673	13
	Australia	0.018	4.818	27

Quercetins		Population		
		p-Value	*F* Value	df Residuals
R. biternatus	across regions	<0.001	17.883	44
	Isthme Bas	0.088	4.697	18
	Australia	**<0.001**	26.440	24

Table 1. *Cont.*

Amines		Population		
		p-Value	*F* Value	df Residuals
R. pseudotrullifolius	across regions	**<0.001**	19.441	46
	Isthme Bas	>0.500	0.202	22
	Australia	**<0.001**	56.282	22
R. moseleyi	across regions	**<0.001**	78.570	36
	Isthme Bas	>0.500	1.075	13
	Australia	**<0.001**	147.176	21

* For each species, ANCOVAs were conducted either across or within regions with "population" as factor and environmental conditions as co-variables. Each line represents a separate ANCOVA and only the result for the factor "population" is shown. *p*-values within regions are sequential Bonferroni's corrected and bold when <0.05.

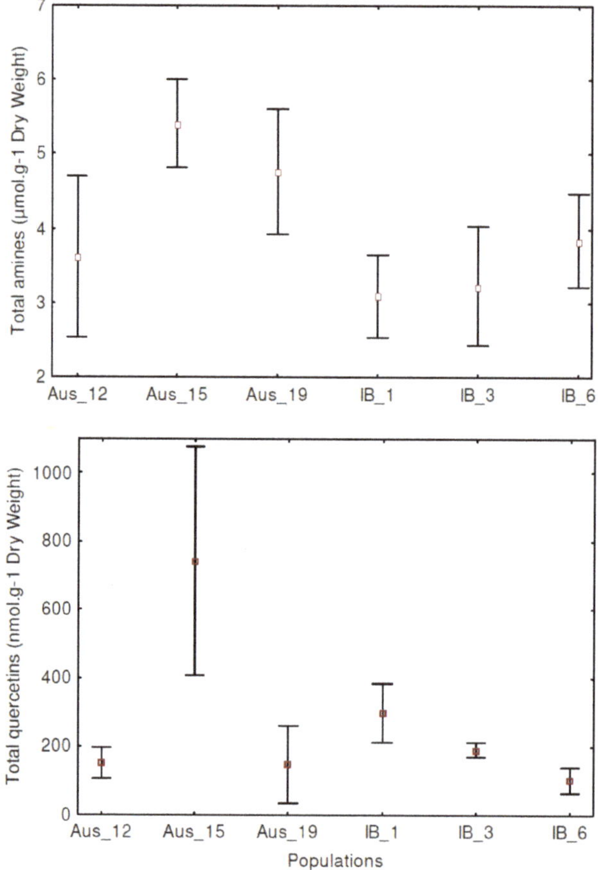

Figure 1. Variance of total contents of amines or quercetins among populations. Means and SD (standard deviation) are given. See Table 1 for test statistics.

To identify an effect of population on compositions (rather than totals) of metabolites, we conducted multivariate ANCOVAS with environmental conditions as co-variables. We found that metabolite (amine or quercetin) composition differed significantly among populations across regions within species, i.e., statistically independent of the environmental variables considered (Table 2).

Table 2. Differences among populations in the composition of amines or quercetins *.

Amines		Population		
		p-Value	*F*-Value	df Residuals
R. biternatus	across regions Isthme Bas	<0.001	5.850	44
	Australia	<0.001	9.554	24
R. pseudotrullifolius	across regions Isthme Bas	<0.001	11.256	48
	Australia	<0.001	7.293	24
R. moseleyi	across regions Isthme Bas	<0.001	11.159	48
	Australia	<0.001	18.628	22

Quercetins		Population		
		p-Value	F Value	df Residuals
R. biternatus	across regions Isthme Bas	<0.001	7.254	44
	Australia	0.001	16.269	24
R. pseudotrullifolius	across regions Isthme Bas	<0.001	9.495	46
	Australia	<0.001	16.336	22
R. moseleyi	across regions Isthme Bas	<0.001	89.107	36
	Australia	<0.001	126.020	21

* For each species, multivariate ANCOVAS with "population" as factor, environmental variables as covariables, and the levels of all amines (top) or all quercetins (bottom) as dependent variables, done either across regions or within the Australia region (analysis was not possible in Isthme Bas due to collinearity of environmental conditions and populations in this region).

2.2. Environments Partly Explain Variation of Total Metabolite Contents and Metabolite Composition across Populations

For total amine content, relationships with environmental conditions (soil water saturation, pH or conductivity) were significant for five out of nine comparisons in *R. biternatus* (across or within regions) and three in both *R. pseudotrullifolius* and *R. moseleyi* (Table S1).

Amine composition was significantly related to overall environment and individual environmental conditions, within or across regions in the three species (Table 3 and Figure 2 for an example of the full multivariate relationships between all environmental variables and all compounds). Testing the relationships between amine composition and individual environmental conditions, we found that individual amine levels were correlated to soil water saturation in 17 out of 48 comparisons (across or within regions) in *R. biternatus*, 16 in *R. pseudotrullifolius*, and six in *R. moseleyi* (Figure S1). Individual amine levels were correlated to soil pH in 11 out of 48 comparisons in *R. biternatus* and *R. pseudotrullifolius* and 15 in *R. moseleyi*. Individual amine levels were correlated to conductivity in 20 out of 48 comparisons in *R. biternatus*, 21 in *R. pseudotrullifolius*, and 17 in *R. moseleyi* (Figure S1).

Table 3. Multivariate relationship between a given environmental variable and compositions of either all amines or all quercetins. Multivariate multiple regression analysis relating a given environmental variable to levels of all compounds *.

Amines		Overall Environment		Water Saturation		pH		Conductivity		
		F Value	*p*-Value	F Value	*p*-Value	F Value	*p*-Value	F Value	*p*-Value	df Residuals
R. biternatus	across regions	4.055	<0.001	5.539	<0.001	3.022	0.003	17.019	<0.001	41
	Isthme Bas	4.383	0.361	9.575	0.249	6.470	0.301	2.660	0.452	11
	Australia	5.777	0.001	15.488	<0.001	2.754	0.036	9.360	<0.001	13
R. pseudotrullifolius	across regions	5.089	<0.001	7.616	<0.001	2.881	0.003	4.837	<0.001	40
	Isthme Bas	4.811	0.008	17.536	<0.001	1.810	0.172	8.924	<0.001	10
	Australia	4.221	0.006	17.765	<0.001	2.866	0.031	4.279	0.006	13

Table 3. *Cont.*

Amines		Overall Environment		Water Saturation		pH		Conductivity		
		F Value	*p*-Value	*F* Value	*p*-Value	*F* Value	*p*-Value	*F* Value	*p*-Value	df Residuals
R. moseleyi	across regions	4.006	**<0.001**	1.186	0.334	4.103	**<0.001**	6.392	**<0.001**	24
	Isthme Bas									
	Australia	11.250	**<0.001**	6.576	**0.002**	7.502	**<0.001**	19.503	**<0.001**	11
Quercetins		Overall Environment								
		F Value	*p*-Value	*F* Value	*p*-Value	*F* Value	*p*-Value	*F* Value	*p*-Value	df Residuals
R. biternatus	across regions	4.067	**<0.001**	2.718	**0.013**	2.709	**0.014**	4.779	**<0.001**	43
	Isthme Bas	4.169	**0.010**	3.691	**0.017**	5.254	**0.004**	3.668	**0.017**	17
	Australia	14.812	**<0.001**	105.210	**<0.001**	7.995	**0.002**	10.625	**<0.001**	17
R. pseudotrullifolius	across regions	1.139	0.357	5.970	**<0.001**	3.896	**0.001**	2.855	**0.009**	47
	Isthme Bas	22.216	**<0.001**	49.230	**<0.001**	8.963	**<0.001**	40.232	**<0.001**	17
	Australia	3.017	**0.022**	1.714	0.158	7.465	**<0.001**	7.088	**<0.001**	21
R. moseleyi	across regions	5.492	**<0.001**	16.485	**<0.001**	10.166	**<0.001**	6.845	**<0.001**	17
	Isthme Bas	126.530	**<0.001**	39.861	**<0.001**	123.590	**<0.001**	3.425	**0.049**	3
	Australia	56.519	**<0.001**	74.673	**<0.001**	40.919	**<0.001**	23.647	**<0.001**	5

* "Overall environment" represents scores along the first axis of an environmental PCA (Principal Component Analysis, Methods). *p*-values are sequential Bonferroni's corrected and bold when < 0.05. Sample size was insufficient for *R moseleyi* in Isthme Bas.

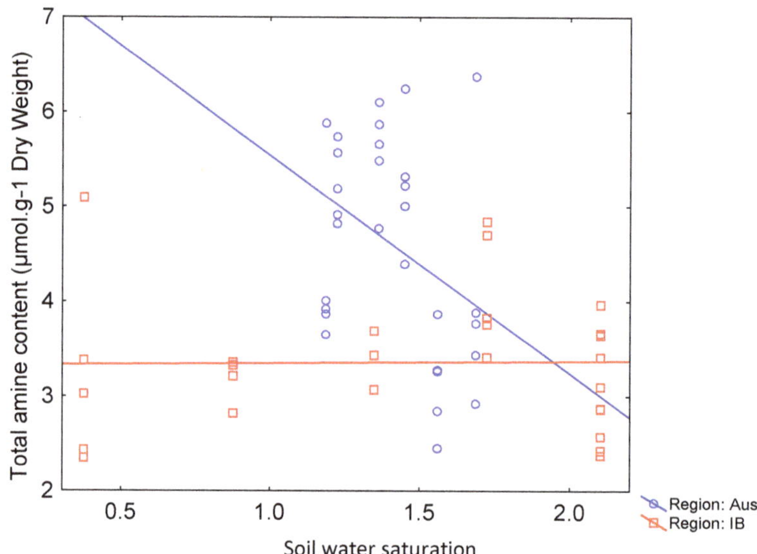

Figure 2. Illustration of region-dependent relationships of total amines to soil water saturation for *R. pseudotrullifolius*.

For total quercetin contents, relationships with environmental conditions were significant for five out of nine comparisons (across or within regions) in either *R. biternatus* or *R. pseudotrullifolius* and 2 in *R. moseleyi* (Table S1).

Quercetin composition was significantly related to overall environment in the three species (Table 3). Individual quercetin levels were significantly correlated to water saturation level in nine out of 27 comparisons (across or within regions) in *R. biternatus*, 11 in *R. pseudotrullifolius*, and 16 in *R. moseleyi* (Figure S2). Individual quercetin levels were significantly correlated to soil pH in 10 out of 27 comparisons in *R. biternatus*, 12 in *R. pseudotrullifolius*, and seven in *R. moseleyi*. Finally, they were significantly correlated to soil conductivity in 13 out of 27 comparisons in *R. biternatus*, eight in *R. pseudotrullifolius*, and six in *R. moseleyi* (Figure S2).

2.3. Morphological Phenotypes Partly Explain Variation of Total Metabolite Contents and Metabolite Composition across Populations

For total amine content, relations with morphological traits were significant in one out of 15 comparisons (across or within regions) in *R. biternatus*, three in *R. pseudotrullifolius*, and four in *R. moseleyi* (Table S2).

Amine composition was significantly related to the morphological phenotype across or within regions in the three species (Table 4). Traits correlated to amine composition were mainly plant height and the number of leaves (Figure S3). Individual amine levels were correlated to plant height in nine out of 48 comparisons (across or within regions) in *R. biternatus*, 8 in *R. pseudotrullifolius*, and 11 in *R. moseleyi*. Individual amine levels were correlated to the number of leaves in four out of 48 comparisons in *R. biternatus*, six in *R. pseudotrullifolius* and four in *R. moseleyi* (Figure S3).

For total quercetin content, relations with phenotypic traits were significant in three out of 15 comparisons (across or within regions) in *R. biternatus*, three in *R. pseudotrullifolius*, and five in *R. moseleyi* (Table S2).

Quercetin composition was significantly related to the morphological phenotype in *R. biternatus*, *R. pseudotrullifolius* and *R. moseleyi* (Table 4). Traits influenced by quercetin composition were mainly plant height and the number of leaves (Figure S4). Individual quercetin levels were significantly correlated to plant height in three out of 27 comparisons in *R. biternatus*, eight in *R. pseudotrullifolius*, and nine in *R. moseleyi*. Individual quercetin levels were significantly correlated to the number of leaves in 13 out of 27 comparisons in *R. biternatus*, nine in *R. pseudotrullifolius*, and one in *R. moseleyi* (Figure S4).

2.4. Regions and Metabolite-Environment Relationships

For total metabolite contents, we found 3/12 (amines) and 4/12 (quercetins) significant effects of regions on the relationships between total metabolite contents and environmental conditions (overall environment, soil water saturation, pH, conductivity) (Table 5, examples in Figure 2). These significant effects concerned *R. pseudotrullifolius* and *R. moseleyi* but not *R. biternatus*.

For amine composition, in multivariate analyses the relationship between environmental conditions and levels of individual amines was significant in one region but not in another in five out of eight comparisons within *R. biternatus* and *R. pseudotrullifolius* (Table 3). Overall, environment explained variance in amine composition better within than across regions in nine out of 12 comparisons (Table 3). In univariate analyses, the relationship between environmental conditions and levels of individual amines was significant in one region but not the other in 57 out of 144 comparisons, was significant in both but changed sign in one comparison, and was significant in both with the same sign in 11 out of 144 comparisons (Figure S1). Changes among regions were two times more frequent for soil water saturation or conductivity than for pH (24/1/1 and 21/0/7 vs. 12/0/3; respectively changes from significant to not significant/positive significance to negative/significant with same sign). Also, changes among regions were two times more frequent in *R. pseudotrullifolius* and *R. biternatus* than in *R. moseleyi* (23/1/3 and 24/0/1 vs. 10/0/7) (Figure S1). For example, in *R. biternatus*, a redundancy analysis accounting simultaneously for all multiple amines showed major shifts of relationships among regions: see, for instance, the changes in relative positions of Agm, Put, Spd and Dop (Figure 3).

Table 4. Multivariate relationship between a given trait and compositions of either all amines or all quercetins. Multivariate multiple regression analysis relating a given trait to levels of all compounds *.

Amines		Phenotype		Plant height		Number of Leaves		Flower Size		Leaf Ratio		LDMC		df Residuals
		p Value	*F*	*p* Value	*F*	*p* Value	*F*	*p* Value	*F*	*p* Value	*F*	*p* Value	*F*	
R. biternatus	across regions	0.150	1.491	0.200	1.379	0.540	0.934	0.260	1.272	0.110	1.609	0.130	1.546	41
	Isthme Bas	**0.006**	4.785	0.220	1.590	0.111	2.079	0.150	1.860	**0.005**	5.001	0.300	1.376	11
	Australia	0.350	1.243	0.570	1.041	0.700	0.763	0.230	1.508	0.560	0.455	0.340	1.262	13
R. pseudotrullifolius	across regions	**0.006**	2.670	0.720	0.758	0.096	1.664	**0.025**	2.154	0.092	1.680	0.075	1.755	40
	Isthme Bas	**0.005**	5.422	0.600	0.886	**0.008**	4.798	**0.027**	3.418	0.510	1.011	**0.006**	5.173	10
	Australia	**0.012**	3.621	0.760	0.692	**0.005**	4.413	**0.050**	2.515	**0.009**	3.871	**0.006**	4.240	13
R. moseleyi	across regions	**0.015**	2.654	0.290	1.271	0.031	2.310	**0.020**	2.516	**0.006**	3.108	0.280	1.288	24
	Isthme Bas													
	Australia	0.088	2.338	**0.006**	5.173	0.140	1.966	0.390	1.209	0.015	4.046	0.280	1.449	11

Quercetins		Phenotype		Plant height		Number of Leaves		Flower Size		Leaf Ratio		LDMC		df Residuals
		p Value	*F*	*p* Value	*F*	*p* Value	*F*	*p* Value	*F*	*p* Value	*F*	*p* Value	*F*	
R. biternatus	across regions	**0.025**	2.435	0.240	1.353	**0.024**	2.454	0.320	1.201	0.380	1.105	0.120	1.698	43
	Isthme Bas	0.475	1.004	0.180	1.661	0.084	2.174	0.690	0.713	0.920	0.395	0.910	0.412	17
	Australia	0.350	1.217	0.480	0.997	0.540	0.910	0.680	0.726	0.130	1.878	0.120	1.932	17
R. pseudotrullifolius	across regions	**0.005**	3.137	**0.006**	3.053	**0.004**	3.239	**0.006**	3.053	**0.024**	2.424	0.520	0.916	47
	Isthme Bas	**0.006**	4.221	0.240	1.470	**0.024**	3.080	**0.010**	3.780	0.190	1.626	0.570	0.869	17
	Australia	**0.006**	3.822	0.120	1.854	0.072	2.166	0.260	1.383	**0.035**	2.619	0.057	2.311	21
R. moseleyi	across regions	**0.006**	4.221	**0.009**	3.869	0.089	2.135	**0.005**	4.384	**0.004**	4.588	0.120	1.932	17
	Isthme Bas	**0.010**	99.388	0.220	3.915	0.183	4.837	**0.006**	166.055	0.056	17.241	0.110	8.471	3
	Australia	0.820	0.503	0.460	1.209	0.342	1.606	0.780	0.567	0.250	2.081	**0.006**	19.212	5

* "Phenotype" represents scores along the first axis of a PCA calculated across the traits. Indicated are *F*- and *p*-values after sequential Bonferroni's correction. *p*-values are sequential Bonferroni's corrected and bold when < 0.05. Sample size was insufficient for *R. moseleyi* in Isthme Bas.

Table 5. Effects of region on the relationships between total metabolite contents (amines or quercetins) and (**a**) the environment, or (**b**) the phenotype among populations. The table shows the interaction term between region and either a given environmental variable or a given trait *.

	Interaction Term between Regions and											
Amines	**(a) Overall Environment**			**Soil Water Saturation**			**pH**			**Conductivity**		
	p-Value	F	R^2	*p*-Value	F	R^2	*p*-Value	F	R^2	*p*-Value	F	R^2
R. biternatus	0.059	3.723	0.213	0.157	2.07	0.116	0.291	1.138	0.007	0.608	0.267	0.52
R. pseudotrullifolius	0.349	0.892	0.306	**0.023**	5.467	0.336	0.952	0.004	0.309	0.116	2.55	0.32
R. moseleyi	0.111	2.647	0.527	**<0.001**	23.706	0.317	0.068	3.522	0.375	**0.051**	4.048	0.55
Quercetins	**Overall Environment**			**Soil Water Saturation**			**pH**			**Conductivity**		
	p-Value	F	R^2	*p*-Value	F	R^2	*p*-Value	F	R^2	*p*-Value	F	R^2
R. biternatus	0.949	0.004	0.21	0.191	1.755	0.196	0.104	2.747	0.356	0.073	3.351	0.18
R. pseudotrullifolius	**0.005**	8.719	0.104	**0.005**	8.719	0.011	**<0.001**	14.71	0.204	0.312	1.044	0.17
R. moseleyi	0.593	0.291	0.006	0.502	0.459	0.314	0.574	0.321	0.062	**0.001**	11.71	0.31

	Interaction Term between Regions and								
Amines	**(b) Phenotype**			**Plant height**			**Number of Leaves**		
	p-Value	F	R^2	*p*-Value	F	R^2	*p*-Value	F	R^2
R. biternatus	0.91	0	0.042	0.89	0.774	0.025	0.094	4.709	0.066
R. pseudotrullifolius	0.99	0.029	0.26	0.31	0.203	0.27	**0.001**	10.82	0.39
R. moseleyi	**0.046**	9.388	0.35	**0.042**	5.76	0.32	0.41	0.048	0.26
Quercetins	**Phenotype**			**Plant height**			**Number of Leaves**		
	p-value	F	R^2	*p*-value	F	R^2	*p*-value	F	R^2
R. biternatus	0.63	0.005	0.008	0.76	0.757	0.022	0.61	0.152	0.17
R. pseudotrullifolius	0.77	0.084	0.025	0.87	0.026	0.1	**0.022**	7.952	0.13
R. moseleyi	0.17	3.349	0.18	0.08	0.078	0.11	0.57	2.624	0.1

* Interaction terms are taken from multiple regression analyses that contain also the respective raw variables. "Overall environment" and "Phenotype" represent scores along the first axis of an environmental or morphological PCA, respectively (Methods) calculated across the "soil water saturation", "pH" and "conductivity". F and sequential-Bonferroni corrected *p*-value (bold if < 0.05) of the interaction term; and adjusted r squared of the multiple regression. Sample sizes for amines/quercetines: *R. biternatus* = 58/52; *R. pseudotrullifolius* = 57/48; *R. moseleyi* = 46/26.

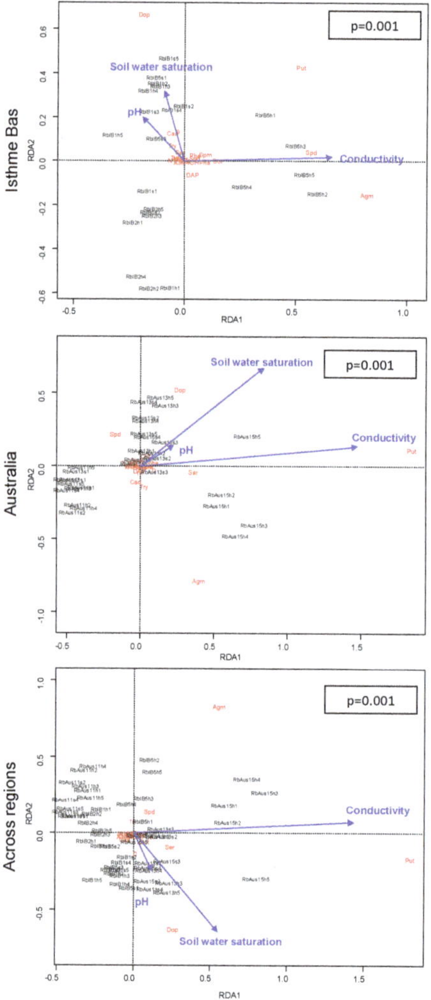

Figure 3. Full relationship between all environmental variables and all amine compounds, within and across regions, exemplified for *R. biternatus*. Environmental variables are soil water saturation, pH and conductivity. In redundancy analyses, axes are constrained to show, at best, variance explained by environmental variables. Individual, amines (in bold) and environmental variables (with arrows) are indicated. N = are 28 in Isthme Bas; 30 in Australia; 58 across regions. *p* values (*p*) are indicated. R^2 are 0.49, 0.73, ad 0.54 for Isthme Bas, Australia, and across-region, respectively. Abbreviation: Agm: Agmatine; Put: Putrescine; Spm: Spermine; Spd: Spermidine; Cad: Cadaverine; DAP: 1,3-diaminopropane; Dop: Dopamine; Ser: Serotonin; Tyr: Tyramine; Oct: Octopamine; N1Ac-Spm: N^1-acetylspermine; N8Ac-Spd: N^8-acetylspermidine; Try: Tryptamine; 3M4OHPhe: 3-methoxy-4-hydroxy phenylethylamine (see Table 3, Figures S1 and S2 for more detailed analysis, for the full set of species, and for both amines and quercetins).

For quercetin composition, in multivariate analyses the relationship between environmental conditions and levels of individual quercetins was significant in one region but not in the other in one out of 12 comparisons (Table 3). Overall, environment explained variance in quercetin composition better within than across regions in nine out of 12 comparisons (Table 3). In univariate analyses, the relationship between environmental conditions and levels of individual quercetins was significant in one region but not the other in 36 out of 81 comparisons, was significant in both but changed sign in two comparisons, and was significant in both with the same sign in eight out of 81 comparisons (Figure S2). Changes among regions were somewhat more frequent with respect to pH than with respect to soil water or conductivity (15/1/1 vs. 10/0/4 and 11/1/3) and had roughly similar relative frequency in all three species (13/0/4; 12/2/1; 11/0/3) (Figure S2).

2.5. Regions and Metabolite-Phenotype Relationships

For total metabolite contents, we found some 3/9 (amines) and 1/9 (quercetins) significant effects of regions on the relationships between total metabolite contents and morphological variables (Table 5). These significant effects concerned *R. pseudotrullifolius* and *R. moseleyi* but not *R. biternatus*.

For amine composition, in multivariate analyses the relationship between morphological variables and levels of individual amines was significant in one region but not in the other in three out of 12 comparisons (Table 4). Overall, morphology explained variance in amine composition better within than across regions in four out of 18 comparisons (Table 4). In univariate analyses, the relationship between morphological variables and levels of individual amines was significant in one region, but not the other in 27 comparisons, and once significant in both regions, but with opposite signs (out of 144 comparisons) (Figure S3). Changes among regions were two times more frequent for plant height than for leaf number (19/1/0 vs. 8/0/0). Changes among regions were somewhat rarer for *R. biternatus* than for either *R. pseudotrullifolius* or *R. moseleyi* (8/0/0, vs. 10/0/0 and 9/1/0) (Figure S3). Even for *R. biternatus*, a redundancy analysis accounting simultaneously for all multiple amines showed that amine/phenotype relationships are significant within one region (Isthme Bas) and non-significant across regions, with major shifts of the relationships among regions (Figure 4).

For quercetin composition, in multivariate analyses the relationship between morphological variables and levels of individual quercetins was significant in one region, but not in the other in two out of 18 comparisons (Table 4). Overall, morphology explained variance in quercetin composition better within than across regions in three out of 18 comparisons (Table 4). In univariate analyses, the relationship between morphological variables and levels of individual quercetins was significant in one region but not the other in 23 comparisons and was never significant in both regions (out of 81 comparisons) (Figure S4). Changes among regions were roughly equally frequent for plant height and for leaf number (11/0/0 vs. 12/0/0). Changes among regions were somewhat at least two times more abundant in *R. biternatus* and *R. pseudotrullifolius* than in *R. moseleyi* (9/0/0, vs. 10/0/0 and 4/1/0) (Figure S4).

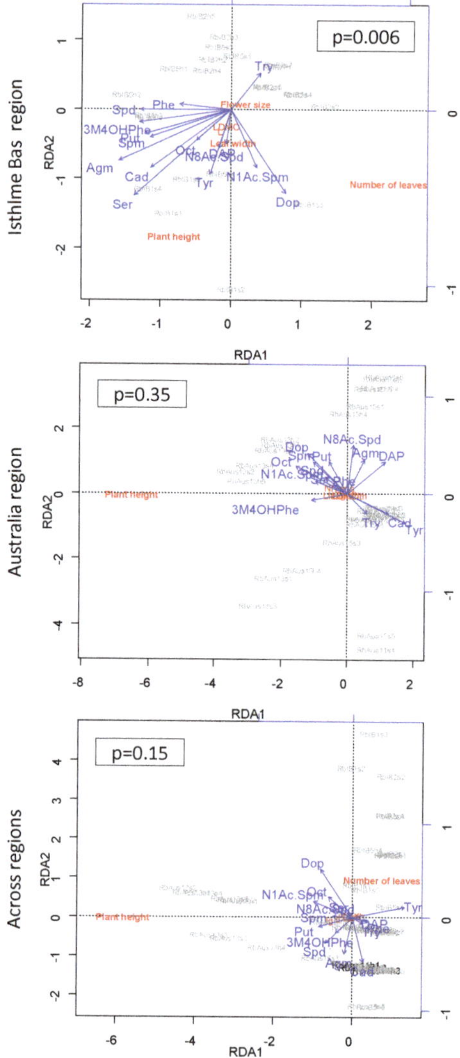

Figure 4. Full relationship between all amine compounds and all traits in *R. biternatus* within and across regions, from redundancy analysis. Traits are Plant height, Number of leaves, Flower size, Leaf length, Leaf width, LDMC (Leaf Dry Matter Content). In redundancy analyses, axes are constrained to show, at best, variance explained by amines. Individual traits (in bold) and amines (with arrows) are indicated. N = are 28 in Isthme Bas; 30 in Australia; 58 across regions. *p* values (*p*) are indicated. Abbreviation: Agm: Agmatine; Put: Putrescine; Spm: Spermine; Spd: Spermidine; Cad: Cadaverine; DAP: 1,3-diaminopropane; Dop: Dopamine; Ser: Serotonin; Tyr: Tyramine; Oct: Octopamine; N1Ac.Spm: N^1-acetylspermine; N8Ac.Spd: N^8-acetylspermidine; Try: Tryptamine; 3M4OHPhe: 3-methoxy-4-hydroxy phenylethylamine (see Table 4, Figures S3 and S4 for more detailed analysis, for the full set of species, and for both amines and quercetins).

3. Discussion

While the macroevolution of secondary metabolites is becoming more and more deciphered, their microevolution at the intraspecific level remains far more obscure [14,25]. We investigated how secondary metabolites vary at the intraspecific level, where plants are subject to both environmental constraints and neutral microevolution. We hypothesized that microevolution within species could affect metabolite composition but also metabolite functions—i.e., the way metabolites respond to the environment or interact with the phenotype. In a correlational multi-species field study, we investigated two secondary metabolite families (amines and quercetins) among natural populations in sub-Antarctic Iles Kerguelen. Populations were distributed across two distinct regions, each region having similar environmental gradients. This pattern allows determining differences between regions in environment-metabolite or trait-metabolite relationships. Amine and quercetin compositions found in the three species were consistent with previous work [26,36]. We know from respectively Hennion et al. [26] and Hennion et al. [36] that amine or quercetin compositions in natural populations of *Ranunculus* species from Iles Kerguelen differ among environments and species. We found that variation of amine or quercetin composition among populations or regions within species is not only shaped by the environment, suggesting that either neutral microevolution or evolution under past environmental selection pressures may also shape metabolite composition within species. Moreover, we found that depending on regions, different metabolites may be correlated to the same environmental condition or the same trait, which is consistent with the hypothesis of metabolite redundancy within species. Also, depending on regions, a given metabolite may show different or even opposite correlations, which supports the hypothesis of metabolite versatility within species. Our results suggest that microevolution within species shapes secondary metabolite composition but also influences metabolite functions through metabolite redundancy and versatility.

3.1. Environments and Morphological Traits Partly Explained Variation of Total Metabolite Contents and Metabolite Composition across Populations

For all three species, the highest number of correlations of individual amine levels was with soil or water conductivity (Table 3 and Figure S1). In *R. biternatus* and *R. pseudotrullifolius* many correlations of individual amine levels were found with soil water saturation, more than with pH, with the reverse case in *R. moseleyi*. These findings are consistent with the known involvement of amines in plant response to water and salinity, some having demonstrated protective roles against these stresses [37]. In contrast, relations with soil pH are less known.

Regarding quercetins, there were about as many correlations of individual quercetin levels with environmental conditions as with amines, invoking equally all three environmental conditions (Table 3 and Figure S2). Quercetins are involved in plant response to stress [35], however, to our knowledge, our study is the first that raised correlations between individual environmental conditions and quercetin composition of plants in nature. Our results suggest a possible involvement of quercetins in plant response to soil water saturation, conductivity and pH.

There were relatively few significant correlations of individual compounds (amines or quercetins) with morphological traits (except for quercetins with the number of leaves) (Table 4, Figures S3 and S4). Whatsoever, our study demonstrated that amine or quercetin compositions are related to both environmental conditions and traits in nature within species, a topic rarely addressed for amines [41] and never yet for quercetins [39].

3.2. Environment Did Not Explain All Differences in Total Metabolite Contents and Metabolite Composition among Populations

Only a few authors differentiate metabolite variation that is environment-dependent and metabolite variation that is environment-independent [13,25,26]. Here, we showed that the different populations within species share the same pool of compounds but differ quantitatively in the total contents and in the composition in amines or quercetins. Moreover, we showed that such differences in amine or quercetin totals and compositions among populations were only partly explained by the environment we had measured (Tables 1 and 2, Figure 1). Our results might possibly suggest that we have left out important environmental variables. However, we consider this unlikely given that our previous work supports the soil hydric variables we used as discriminant for the compositions in respectively amines [26] and quercetins [26,36] of our study species. More likely, our results suggest that, to some degree, present differentiation of amines and quercetins within plant species is shaped either by adaptation to unknown past environments (not correlated to present environments), or by neutral microevolutionary differentiation among populations.

3.3. Differences in Relationship of Metabolites with Environment or Traits between Regions; Patterns Consistent with Metabolite Redundancy

Often, in the two different regions, the same environmental condition correlated to different compounds among amines or among quercetins. However, this did not apply to the amine composition in *R. moseleyi*: in this species, the amines that correlated to pH or conductivity were the same (four compounds) across the two regions (Figure S1).

The change in correlation significance from one region to the other was also observed for morphological traits, however here, the numbers of correlations were far less than with individual environmental conditions and mainly concerned one region (Isthme Bas) (Figures S3 and S4). As a result, there were fewer examples of "alternative" compounds correlated to the same trait across the two regions than there were when considering environmental conditions.

Whereas all three *Ranunculus* species studied displayed changes of correlations between the two regions, some differences were observed among the species. *R. pseudotrullifolius* and *R. moseleyi* showed effects of regions on the relations between the total contents of amines or quercetins and environments or traits, but not *R. biternatus* (Table 5). In contrast, at least for environmental conditions, changes in correlations of individual amine or quercetin levels with individual conditions were more frequent in *R. biternatus* and *R. pseudotrullifolius* than in *R. moseleyi* (Table 3, Figure 3, Figure S1 and S2) (whereas correlations with traits were few—Table 4, Figure 4, Figure S3 and S4). The latter result may link to the fact that the changes in correlations among regions most frequently concerned soil water saturation and conductivity, two factors less variable in the aquatic biotopes of *R. moseleyi*. Clearly, the species-specific variation patterns deserve further investigation.

Amines are known to respond to environmental stresses (reviewed in [37]). Such is also true of quercetins [35]. So, raising correlations of these compounds to environmental conditions is not surprising. In contrast, the finding that the same environmental condition correlates to different compounds across different regions is, to our knowledge, reported for the first time. If we consider the studied compounds as functional, it means that the same function in plants growing in different regions may be ensured by different metabolites. This is the definition of functional redundancy. We found this pattern in two metabolite families (amines and quercetins), and we found it among metabolites within species. Such possible metabolite redundancy among plants within species has been studied very little so far [28]. This concept has been initially defined at the interspecific level. Groppa and Benavides [3] reported that across studies different secondary metabolites responded to the same environments among different species—even if many authors emphasized that different studies are often hardly comparable.

Within each of the amine and quercetin families, the compounds share metabolic pathways and may derive one from each other [7,39]. Moreover, metabolites belonging to the same family often show similar, or at least overlapping, functions in plants [7,37,39]. The proximity in synthesis pathway might trigger redundancy in function in respective amines and quercetins.

Our results strongly suggest the occurrence of functional redundancy of metabolites within species. We see an interesting issue that deserves further investigation. All the metabolites are found in all the populations studied across regions: hence, differences in metabolite-environment or metabolite-trait relationships between regions are not due to the presence or absence of different metabolites between regions. In contrast, each population potentially synthesizes several compounds that may not be used to respond to environment or interact with traits. Producing unused compounds may have a cost to plants and costs and benefits of metabolite redundancy remains to be understood.

3.4. Differences in Relationship of Metabolites with Environment or Traits between Regions; Patterns Consistent with Metabolite Versatility

In some cases, the same compound related to different environmental variables or morphological traits in different regions, or a given relationship changed sign (Figures S1–S4). Secondary metabolites may be involved in different functions in different organs or in different environments within the same plant species, i.e., metabolite versatility [5,31,32]. Here we expand the concept of metabolite versatility by suggesting that the same metabolites may have different functions, and even have opposite functions facing the same environment, in different regions. Due to different or opposite effects of the same metabolite in different regions, an analysis across regions comes to the false conclusion that this metabolite likely has no effect at all. Overall, we suggest to take into account versatility in analyzing metabolite functions. The capacity of a metabolite to have different roles increases the capacity of a species to respond to the environment.

3.5. Metabolite Redundancy and Versatility as a Result of Microevolution Driven by Distance rather than Environment

In both regions we sampled roughly the same environmental gradients. Hence, we may probably exclude environmental differences as a cause of pattern of metabolite redundancy and versatility. Alternatively, such patterns might reflect either different metabolite plasticity or different heritable adaptations among populations within the different regions. We do have indication from previous work that some part of the amine composition of plants in natural populations is heritable or persistent over a few years [26,37].

If partly heritable metabolites relate differently to the same environments or traits in two spatially distant but environmentally similar regions, this suggests neutral microevolution between regions. In theory, such differences might also reflect adaptive microevolution under selection pressures relating to past environmental conditions, although general environmental conditions have persisted in Kerguelen since the last glacial maximum [42]. Future research should investigate whether differences of metabolite/environment/phenotype relationships between regions are heritable. Note that the three studied *Ranunculus* species exhibit frequent vegetative reproduction [43] and self-pollination, with even cleistogamy in submerged flowers in *R. pseudotrullifolius* and *R. moseleyi* [44]. Self-pollination and vegetative reproduction within populations may reduce gene flows across populations and may therefore increase microevolutionary differentiation among populations, possibly facilitating the origin of redundancy and versatility.

4. Materials and Methods

4.1. Plant Collection

The Iles Kerguelen (49°20′00″ S, 69°20′00″ E) are located in the Southern Indian Ocean within the sub-Antarctic region [45] (Figure 5). These islands are characterized by permanently low temperature (4.6 °C annual mean), strong and permanent winds (10 m·s^{-1} annual mean), and high precipitation (60 years ago but drastically reduced in recent years) (annual mean of 760 mm in the studied regions) [46]. We studied three *Ranunculus* species (i.e., *R. biternatus*, *R. pseudotrullifolius* and *R. moseleyi*), of which *R. biternatus* is austral circumpolar, *R. pseudotrullifolius* magellanic and on Kerguelen, and *R. moseleyi* is a strict Kerguelen endemic [47]. All are perennial plants. On the Iles Kerguelen, these species have different ecological amplitudes but occupy partially overlapping habitats [43]. *Ranunculus biternatus* is widespread on the island occurring in habitats up to 500 m above sea level. *R. pseudotrullifolius* and *R. moseleyi* have more restricted distributions. The first one, being halophilous, occurs within a short distance of the coast, occupying peaty or sandy shorelines and ponds [43]. *Ranunculus moseleyi* is strictly aquatic, growing only in freshwater lakes and ponds [43].

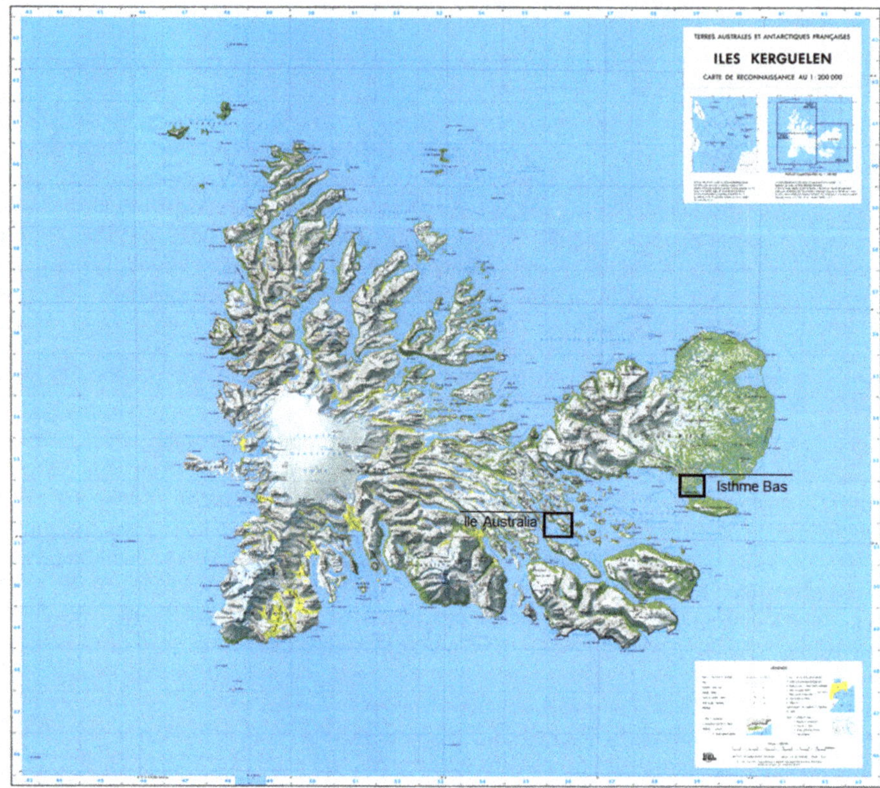

Figure 5. Sampled regions in Iles Kerguelen (modified from IGN [Institut Géographique National, France] map).

Plants were sampled in 18 populations (six populations per species) equally distributed across two regions in Iles Kerguelen: Isthme Bas, a large flat isthmus (about 30 km^2) and Ile Australia, a large island (about 20 km^2) (Figure 3). Plants across the two regions were sampled at similar altitudes and subject to similar temperatures and across the same vegetation types (herbfields, shoreline and pond vegetation) [43]. Each population was subdivided into two sites of contrasting humidity conditions and in each a continuous group of plants living in a same site was sampled. Sampling was performed across a small area (between 0.5 and 5 square meters) so as to ensure that plants assigned to a same population were subject to roughly similar environmental conditions.

The entire sampling was performed in summer during a short period, 6 weeks from mid-December 2011 to mid-March 2012. Plants were sampled between 11 h and 17 h to avoid bias from daily variation of metabolism [48,49]. An average of five individual plants per population, of the most frequent size in the local population were sampled. As the three *Ranunculus* species propagate vegetatively via runners, to avoid pseudoreplication we sampled plants at a distance above average runner length one from each other [50]. In each individual we performed morphological measurements and collected two to four leaves to quantify amines and quercetins. As metabolite composition may vary within a given plant at a given moment between leaves of different developmental stages [49,51], we sampled an appropriate and constant leaf developmental stage, i.e., fully developed photosynthetic leaves as in previous work [52]. We moreover verified whether inclusion of sampling month into an ANOVA relating region to the different compounds reduces the number of significant relationships between region and compounds and found that this is not the case. For amines, even the identity of the affected compounds remained the same in 15 out of 20 cases. The samples collected were frozen in liquid nitrogen and stored at −80 °C then lyophilized and ground to powder using a mixer mill MM 400 Retsch (Haan, Germany).

4.2. Plant Measurements

In each individual we measured plant height, the length and width of the largest leaf, the numbers of leaves and flowers, the flowering stage and the size of the largest flower (largest diameter). Flowering stage of the individuals was estimated following Hennion et al. [26]. To determine leaf dry matter content (LDMC), in each population a total of 20 leaves were sampled from a minimum of 15 individuals and processed following Cornelissen et al. [53]. Leaves were collected then directly put in distilled water in airtight bags for rehydration. They were then weighed before and after 48 h drying at 80 °C. Some traits (flowering stage, number of flowers and the size of largest flower) were highly redundant. We thus only kept the continuously varying trait "size of largest flower" as floral trait. Likewise, we found low variability of LDMC among populations; hence we did not use this trait in our analysis.

4.3. Determination and Quantitation of Free Amines and Acetylated Polyamines

This determination followed Hennion et al. [26], with modified quantities as follows. Samples from individual plants were analyzed individually. Several samples from the same population and environmental conditions were pooled in case of insufficient material (see Table S3). Five to ten milligrams of powdered samples were thoroughly mixed with 100 to 200 µL of 1 mmol·L^{-1} HCl supplemented with 10 µmol·L^{-1} diaminoheptane (Sigma, St. Louis, MO, USA), as an internal standard, on a magnetic stirring plate (2000 rpm) for 1 h at 4 °C. The homogenates were then centrifuged for 15 min at 10,000× g at 4 °C, and the supernatant of each sample collected. The pellet of each sample was further extracted twice with 100 to 200 µL of 1 mmol·L^{-1} HCL and 10 µmol·L^{-1} diaminoheptane. After a short stirring period, the homogenates were centrifuged for 15 min at 10,000× g at 4 °C. For each sample, the three supernatants were combined and used as the crude extracts for characterization and determination of free and acetylated amines and polyamines and stored frozen at −20 °C before chromatographic analyses. High Performance Liquid Chromatography (HPLC) and fluorescence spectrophotometry were used to separate and quantify amines prepared as their dansyl derivatives according to Smith and Davies [54] with some modifications as follows. Aliquots (200 µL) of the supernatant were added to 200 µL of saturated sodium carbonate and 600 µL of dansyl chloride in acetone (7.5 mg·mL^{-1}) in a 5 mL tapered reaction vial. After a brief vortexing, the mixture was incubated in darkness at room temperature for 16 h. Excess dansyl chloride was converted to dansylproline by 30 min incubation after adding 300 µL (150 mg·mL^{-1}) of proline. Dansylated amines were extracted in 1 mL ethylacetate. The organic phase was collected then evaporated to dryness, and the residue was dissolved in methanol and stored in glass vials at −20 °C. External standards were made for agmatine (Agm), diaminopropane (DAP), putrescine (Put), cadaverine (Cad), spermidine (Spd), spermine (Spm), N-acetylputrescine (NAc-Put), N^8-acetylspermidine (N^8Ac-Spd), and N^1-acetylspermine (N^1Ac-Spm) for aliphatic amines and their acetylated conjugates; phenylethylamine (Phe), octopamine (Oct), 3-methoxy-4-hydroxy phenylethylamine (3M4OHPhe), tyramine (Tyr), and dopamine (Dop) for phenylalkylamines; tryptamine (Try) and serotonin (Ser) for indolalkylamines (all authentic products from Sigma, St. Louis, MO, USA). These standards were processed in the same way as samples, and 2 to 50 nmol (per assay) were dansylated for each standard alone or in combination. One standard combined these 15 amines plus diaminoheptane. The HPLC column was packed with reverse-phase SpherisorbODS2 C18 (particle size 5 µm; 4.6 × 250 mm, Waters, Milford, CT, USA). The mobile phase consisted of a solution of 17.5 mmol·L^{-1} potassium acetate (pH 7.17) as eluent A and acetonitrile as eluent B. The solvent gradient, modified according to Hayman et al. [55] was as described by Jubault et al. [56]. The flow rate of the mobile phase was 1.5 mL·min^{-1}. For fluorescence detection of dansyl amines, an excitation wavelength of 366 nm was used with an emission wavelength of 490 nm. The external standards were injected in the HPLC system first to determine retention times of the various amines on the column and secondly to make calibration curves for quantitation. Peaks of amines in the samples were determined by their retention times on the column, and stability was checked by injection of the combined 16-amine standard in the system every 15 samples. In case of doubt, identities of peaks of amines were confirmed by spiking the sample with known amounts of the authentic standards. Amines in the samples were quantified after yield correction with the internal standard and calibration with the external standards. The stability of quantitative calibration was checked by injection of a Put standard every 10 samples. The HPLC design consisted of a thermoelectron pump (SpectraSystem P1000 XR, Thermo Fisher, San Jose, CA, USA) and (Spectra-Series AS100) autosampler with a 20 µL injection loop, and detection through an FP-2020 Plus fluorometer (Jasco, Inc., Easton, MD, USA). Signals were computed and analyzed using Azur software (Datalys, St Martin d'Hères, France).

4.4. Determination and Quantitation of Quercetins

Quercetins were the sole flavonols detected in *Ranunculus* species from Iles Kerguelen [36]. Samples from individual plants were analyzed individually. Several samples from the same population and environmental conditions were pooled in case of insufficient material (see Table S4). We weighed about 10 mg of plant powder in an Eppendorf tube, and added 1 mL of methanol acidified with 1% formic acid. The tube was vortex-agitated first and put in ultra-sonic bath for 5 min. The tube was then centrifuged briefly and 900 µL of the supernatant was removed using a 1 mL plastic syringe, and filtered using a PTFE 13 mm 0.45 µm syringe filter. The methanol extract was then poured in an injection vial for Ultra Performance Liquid Chromatography (UPLC) analysis; 2 µL of the extract were injected in the Waters UPLC_PDA_ESI_TQD system for flavonol quantitation. The reversed phase column, an Acquity Waters C18 BEH (2.1 × 150 mm) 1.7 µm, was maintained at 30 °C. The solvents used for the binary gradient were A: ultra-pure water with 0.1% formic acid, B: acetonitrile with 0.1% formic, the flow was 0.4 mL/min. The gradient applied was 98% A from 0 to 0.2 min, 10% A from 0.2 to 14 min, 14 to 15 min isocratic 10% A, 15 min to 17 min 98% A, 17 to 20 min isocratic 98% A. The photo diode array detector scanned from 190 to 600 nm and flavonols were detected at 350 nm, external quantitation with some flavonol standards was applied. The UPLC-photodiode array-electrospray-triple quadrupole analytical system allows us to detect compounds for which the molecular ion produced in the electrospray source is in accordance with the molecular structure searched. The capillary voltage was 2.9 kV, the cone voltage was 37 V, the source temperature was maintained at 150 °C and the desolvatation temperature at 400 °C, the desolvatation gas flow was 800 L/h. On the basis of data from the literature on Antarctic *Ranunculus* flavonols, Gluchoff-Fiasson et al. [36], the mass spectrometer detector was programmed to focus on characteristic m/z of those flavonols yet identified. In negative mode, the ions, monitored by the Select Ion Recording method (SIR) for quantification were: 787 (Quercetin-tri-glucoside), 949 (Quercetin-caffeoyl-tri-glucoside), 933 (Quercetin-feruloyl-di-glucoside-pentoside), 919 (Quercetin-caffeoyl di-glucoside-pentoside), 757 (Quercetin-di-glucoside-pentoside), 595 (Quercetin-glucoside-pentoside), 625 (Quercetin-di-glucoside), 963 (Quercetin-feruloyl-tri-glucoside) and the standards 463 (isoquercitrin), 609 (rutin). Depending on compound structure, isoquercitrin or rutin were used as standards for the quantification on the mass spectrometer triple quadrupole detector, the external standard calibration was made daily and linear regression factors were at least 0.99.

4.5. Amines Characterized

We characterized 15 different amines which belonged to four biochemical categories: aliphatic amines and their acetylated conjugates, phenylalkylamines and indolalkylamines. The detected aliphatic amines were: agmatine (Agm), diaminopropane (DAP), putrescine (Put), cadaverine (Cad), spermidine (Spd), spermine (Spm), N^8-acetylspermidine (N^8Ac-Spd) and N^1-acetylspermine (N^1Ac-Spm). Phenylalkylamines were phenylethylamine (Phe), octopamine (Oct), 3-methoxy-4-hydroxy phenylethylamine (3M4OHPhe), tyramine (Tyr), and dopamine (Dop). Indolalkylamines were tryptamine (Try) and serotonin (Ser). The raw data of individual compositions are shown in Table S3. All 15 compounds described were present in the three species, in the two regions. Thus, differences in amine composition between species or regions reflected shifts in levels and not qualitative differences.

4.6. Quercetins Characterized

Following Hennion et al. [36], we performed an analysis of flavonols. Quercetins characterized were: quercetin 3-diglucoside-7-glucoside (Q-3GL), quercetin 3-(caffeyl-glucosyl)glucoside-7- glucoside (Q-3GL+Caf), quercetin 3-(ferulyl-glucosyl)glucoside-7-glucoside (Q-3GL+Fer), quercetin 3-(caffeyl-xylosyl)glucoside-7-glucoside (Q-2GL+Xyl+Caf), quercetin 3-(ferulyl- xylosyl)glucoside-7-glucoside (Q-2GL+Xyl+Fer), quercetin 3-xylosylglucoside-7-glucoside (Q-2GL+Xyl), quercetin 3-xylosylglucoside (Q-GL+Xyl), quercetin 3-diglucoside (Q-2GL) and isoquercitrin (IQC). The raw data of individual compositions are shown in Table S4. All nine compounds described were present in the three species, in the two regions. Thus, differences in quercetin composition between species or regions reflected shifts in levels and not qualitative differences.

4.7. Environmental Measurements

In each population we measured soil water saturation, pH and conductivity. Three samples of soil, each of 20 mL, were collected at the rhizosphere level of the measured plants. To determine soil water saturation, half of each soil sample was dried at 105 °C during 48 h and weighed before and after drying [50]. Soil water saturation in each population was calculated as following: soil water saturation = (soil weight before drying – soil weight after drying) / sol weight after drying. Data were transformed following f(x)=log(x) to reduce positive skewness of the data distribution. The remaining soil was mixed with known volume of distilled water and was then left 18 to 24 h to permit sedimentation of soil particles. Immediately after, pH was determined using a pH meter (BASIC 20 PLUS CRISON, resolution 0.01 pH). After another 18 to 24 h sedimentation, conductivity was determined using a conductivity meter (CONSORT K810, resolution 0.1 μS cm^{-1}) [50].

4.8. Statistical Analyses

To determine differences of *total contents* of metabolites (amines or quercetins) among populations that are not due to differences in the environmental variables, we conducted ANCOVA analyses with the environment as a co-variable. We analyzed the difference between populations, accounting for environmental conditions: Total amine content in sample = pH (continuous) + Conductivity (continuous) + Water saturation (continuous) + population (six categories) + error. We consistently repeated this procedure across all populations and across only the populations within a given region to explore whether populations are more different in one region or another or whether populations vary only across regions, and not between. For this and all following analyses we used R 3.5.0 software [57]. To determine differences of total metabolite contents (amines or quercetins) among populations that may be due to differences in the environmental variables we measured, we regressed total metabolite levels against environmental predictors. We repeated this analysis separately within each of the two regions. Equally, to determine differences of total metabolite contents (amines or quercetins) among populations that relate to morphological phenotypes, we regressed total metabolite levels against morphological predictors. We repeated this analysis separately within each of the two regions.

To determine the relationships between *compositions* of metabolites (amines or flavonoids) and environment or phenotype we performed redundancy analyses, using cca function [58,59] in R 3.5.0 software [57]. We determined relationships between metabolite composition and individual environmental factors or individual traits using redundancy analyses with rda function. Also, we determined relationships between metabolite composition and the overall environment (i.e., taking into account all the environmental factors) or the overall phenotype (i.e., taking into account all the traits) using redundancy analyses. We repeated this analysis separately within each of the two regions.

To determine whether relationships between *compositions* of metabolites and the environment or metabolite-phenotype relationships differ between regions, we conducted multiple regression analyses. Dependent variables were metabolite compounds (either amines or quercetins) and independent variables were either of the environmental variables, region and the interaction term between both. We also used an integrative "overall environment" variable calculated as the scores along the first axis of a PCA across the individual environmental variables. This axis was most strongly correlated to water saturation in *R. biternatus*, and to pH in the two other species. To statistically test whether relationships between *compositions* of metabolites and the morphological phenotypes differ between regions, we took the same approach as for environment, replacing environmental by morphological variables. Again, we identified an integrative "overall phenotype" variable calculated as the scores along the first axis of a PCA across the individual morphological variables. This axis was most strongly correlated to plant height in all three species and to leaf number in *R. biternatus*, and to LDMC in the other two species.

For multiple testing on the same data set, *p*-values were corrected using sequential Bonferroni's correction [60]. In all regression analyses we verified the assumptions of the analyses using QQ plots and predicted-vs-residual plots.

5. Conclusions

Recent authors [13,14] encouraged researchers to explore the relative contributions of genetic, environmental, microenvironmental and stochastic variation to secondary metabolite variation across plant taxa and environments. Our study provides several hints into secondary metabolite variation and microevolution within species. For two metabolite classes, we showed that variation of secondary metabolite composition among populations was only partly related to environment, suggesting that neutral microevolution also shapes metabolite composition within species. We showed differences in metabolite-environment and metabolite-trait relations among regions. The observed variation patterns may be interpreted as metabolite redundancy and versatility within species. Our results suggest that such possible metabolite redundancy and versatility may be shaped by neutral microevolution. Metabolite redundancy and versatility within species may contribute to the high functional diversity of individual secondary metabolites. We found patterns suggesting metabolite redundancy and versatility in three species and in two distinct families of secondary metabolites (i.e., amines and quercetins), relating to all environmental parameters and all morphological variables. Therefore, our observations likely do not result from a special case but may be extendable to other species, secondary metabolite families or locations. Future aims may be to assess the extent of metabolite redundancy and versatility, test the functions of metabolites in nature, and look at these processes in the light of costs and benefits for plant species.

Supplementary Materials: The following are available online at http://www.mdpi.com/2223-7747/8/7/234/s1, Figure S1: relationships between individual amines and individual environmental variables from simple regression analyses, within and across regions, Figure S2: relationships between individual quercetins and individual environmental variables from simple regression analyses, within and across regions, Figure S3: relationships between individual amines and individual traits from simple regression analyses, within and across regions, Figure S4: relationships between individual quercetins and individual traits from simple regression analyses, within and across regions, Table S1: simple regression analyses to describe the relationship between total levels of amines or quercetins and environmental conditions, Table S2: simple regression analyses to describe the relationship between total levels of amines or quercetins and traits. Table S3: sample distribution by population and region, collection dates, and amine composition. Table S4: sample distribution by population and region, collection dates, and quercetin composition.

Author Contributions: Conceptualization, F.H., A.P. and B.L.; methodology, F.H., A.P. and B.L.; validation, F.H., A.P. and B.L.; formal analysis, B.L., A.P. and T.D.; investigation, B.L., E.C. and F.H.; resources, F.H.; data curation, F.H., B.L., E.C. and T.D.; writing—original draft preparation, B.L., F.H. and A.P.; writing—review and editing, F.H. and A.P.; visualization, B.L., E.C. and T.D.; supervision, F.H. and A.P.; project administration, F.H.; funding acquisition, F.H.

Funding: This research was funded by INSTITUT POLAIRE FRANÇAIS IPEV, grants 1116 (PlantEvol) and 136 (Ecobio). B.L. was supported by a PhD grant from Ministry of Research and Education (France). The APC was funded by CNRS LIA "AntarctPlantAdapt".

Acknowledgments: We thank Marc Lebouvier (UMR ECOBIO, Rennes, France), volunteers Marine Pouvreau, Françoise Cardou, Julie Vingère and Marion Lombard (IPEV N°136), Richard Winkworth (IFS, Massey University, New Zealand), Philippe Choler (LECA, Grenoble, France), the IPEV logistics team, and Réserve Naturelle TAAF for help during the 2011–2012 summer campaign in Iles Kerguelen, and Françoise Binet (UMR ECOBIO, Rennes, France) for support to the programme. We thank Nathalie Marnet who performed the flavonoid analyses at P2M2 facility (INRA, Le Rheu, France). We thank Jacqui Shykoff (CNRS, Université Paris-Saclay, Orsay) for helpful comments on an initial version of the manuscript. This research is linked to CNRS Zone-Atelier Antarctique, to CNRS LIA "AntarctPlantAdapt" (F.H.) with New Zealand, and the Scientific Committee on Antarctic Research programmes AntEco and AnT-ERA.

Conflicts of Interest: The authors declare no conflict of interest. The funders had no role in the design of the study; in the collection, analyses, or interpretation of data; in the writing of the manuscript, or in the decision to publish the results.

References

1. Wink, M. Evolution of secondary metabolites from an ecological and molecular phylogenetic perspective. *Phytochemistry* **2003**, *64*, 3–19. [CrossRef]
2. Croteau, R.; Kutchan, T.M.; Lewis, N.G. Natural products (secondary metabolites). *Biochem. Mol. Biol. Plants* **2000**, *24*, 1250–1319.
3. Groppa, M.D.; Benavides, M.P. Polyamines and abiotic stress: Recent advances. *Amino Acids* **2008**, *34*, 35–45. [CrossRef] [PubMed]
4. Wahid, A.; Ghazanfar, A. Possible involvement of some secondary metabolites in salt tolerance of sugarcane. *J. Plant Physiol.* **2006**, *163*, 723–730. [CrossRef] [PubMed]
5. Wink, M. Evolution of secondary metabolites in legumes (Fabaceae). *S. Afr. J. Bot.* **2013**, *89*, 164–175. [CrossRef]
6. Ober, D. Seeing double: Gene duplication and diversification in plant secondary metabolism. *Trends Plant Sci.* **2005**, *10*, 444–449. [CrossRef] [PubMed]
7. Tiburcio, A.F.; Altabella, T.; Bitrian, M.; Alcazar, R. The roles of polyamines during the lifespan of plants: From development to stress. *Planta* **2014**, *240*, 1–18. [CrossRef]
8. Hartmann, T. From waste products to ecochemicals: Fifty years research of plant secondary metabolism. *Phytochemistry* **2007**, *68*, 2831–2846. [CrossRef]
9. Tohge, T.; Watanabe, M.; Hoefgen, R.; Fernie, A.R. The evolution of phenylpropanoid metabolism in the green lineage. *Crit. Rev. Biochem. Mol. Biol.* **2013**, *48*, 123–152. [CrossRef]
10. Pathania, S.; Bagler, G.; Ahuja, P.S. Differential Network Analysis Reveals Evolutionary Complexity in Secondary Metabolism of Rauvolfia serpentina over Catharanthus roseus. *Front. Plant Sci.* **2016**, *7*, 17. [CrossRef]
11. Peng, M.; Gao, Y.Q.; Chen, W.; Wang, W.S.; Shen, S.Q.; Shi, J.; Wang, C.; Zhang, Y.; Zou, L.; Wang, S.C.; et al. Evolutionarily Distinct BAHD N-Acyltransferases Are Responsible for Natural Variation of Aromatic Amine Conjugates in Rice. *Plant Cell* **2016**, *28*, 1533–1550. [CrossRef]
12. Xie, L.L.; Liu, P.L.; Zhu, Z.X.; Zhang, S.F.; Zhang, S.J.; Li, F.; Zhang, H.; Li, G.L.; Wei, Y.X.; Sun, R.F. Phylogeny and Expression Analyses Reveal Important Roles for Plant PKS III Family during the Conquest of Land by Plants and Angiosperm Diversification. *Front. Plant Sci.* **2016**, *7*, 15. [CrossRef]
13. Bundy, J.G.; Davey, M.P.; Viant, M.R. Environmental metabolomics: A critical review and future perspectives. *Metabolomics* **2009**, *5*, 3–21. [CrossRef]
14. Moore, B.D.; Andrew, R.L.; Kulheim, C.; Foley, W.J. Explaining intraspecific diversity in plant secondary metabolites in an ecological context. *New Phytol.* **2014**, *201*, 733–750. [CrossRef]
15. Arany, A.M.; de Jong, T.J.; Kim, H.K.; van Dam, N.M.; Choi, Y.H.; Verpoorte, R.; van der Meijden, E. Glucosinolates and other metabolites in the leaves of Arabidopsis thaliana from natural populations and their effects on a generalist and a specialist herbivore. *Chemoecology* **2008**, *18*, 65–71. [CrossRef]
16. Benetis, R.; Radusiene, J.; Janulis, V. Variability of phenolic compounds in flowers of Achillea millefolium wild populations in Lithuania. *Med. Lith.* **2008**, *44*, 775–781. [CrossRef]
17. Davey, M.P.; Burrell, M.M.; Woodward, F.I.; Quick, W.P. Population-specific metabolic phenotypes of Arabidopsis lyrata ssp petraea. *New Phytol.* **2008**, *177*, 380–388. [CrossRef]
18. Kim, J.; Kang, K.; Gonzales-Vigil, E.; Shi, F.; Jones, A.D.; Barry, C.S.; Last, R.L. Striking Natural Diversity in Glandular Trichome Acylsugar Composition Is Shaped by Variation at the Acyltransferase2 Locus in the Wild Tomato Solanum habrochaites. *Plant Physiol.* **2012**, *160*, 1854–1870. [CrossRef]

19. Nkomo, M.M.; Katerere, D.D.R.; Vismer, H.H.F.; Cruz, T.T.; Balayssac, S.S.; Malet-Martino, M.M.; Makunga, N.N.P. Fusarium inhibition by wild populations of the medicinal plant Salvia africana-lutea L. linked to metabolomic profiling. *BMC Complement. Altern. Med.* **2014**, *14*, 9. [CrossRef]

20. El-Bakry, A.A.; Hammad, I.A.; Rafat, F.A. Polymorphism in Calotropis procera: Preliminary genetic variation in plants from different phytogeographical regions of Egypt. *Rend. Lincei-Sci. Fis. E Nat.* **2014**, *25*, 471–477. [CrossRef]

21. Jandova, K.; Dostal, P.; Cajthaml, T.; Kamenik, Z. Intraspecific variability in allelopathy of Heracleum mantegazzianum is linked to the metabolic profile of root exudates. *Ann. Bot.* **2015**, *115*, 821–831. [CrossRef]

22. Nazem, V.; Sabzalian, M.R.; Saeidi, G.; Rahimmalek, M. Essential oil yield and composition and secondary metabolites in self- and open-pollinated populations of mint (Mentha spp.). *Ind. Crop. Prod.* **2019**, *130*, 332–340. [CrossRef]

23. Demasi, S.; Caser, M.; Lonati, M.; Cioni, P.L.; Pistelli, L.; Najar, B.; Scariot, V. Latitude and Altitude Influence Secondary Metabolite Production in Peripheral Alpine Populations of the Mediterranean Species Lavandula angustifolia Mill. *Front. Plant Sci.* **2018**, *9*, 983. [CrossRef]

24. Iwanycki Ahlstrand, N.; Reghev, N.H.; Markussen, B.; Hansen, H.C.B.; Eiriksson, F.F.; Thorsteinsdottir, M.; Ronsted, N.; Barnes, C.J. Untargeted metabolic profiling reveals geography as the strongest predictor of metabolic phenotypes of a cosmopolitan weed. *Ecol. Evol.* **2018**, *8*, 6812–6826. [CrossRef]

25. Davey, M.P.; Woodward, F.I.; Quick, W.P. Intraspecific variation in cold-temperature metabolic phenotypes of Arabidopsis lyrata ssp petraea. *Metabolomics* **2009**, *5*, 138–149. [CrossRef]

26. Hennion, F.; Bouchereau, A.; Gauthier, C.; Hermant, M.; Vernon, P.; Prinzing, A. Variation in amine composition in plant species: How it integrates macroevolutionary and environmental signals. *Am. J. Bot.* **2012**, *99*, 36–45. [CrossRef]

27. Taft, S.; Najar, A.; Godbout, J.; Bousquet, J.; Erbilgin, N. Variations in foliar monoterpenes across the range of jack pine reveal three widespread chemotypes: Implications to host expansion of invasive mountain pine beetle. *Front. Plant Sci.* **2015**, *6*, 12. [CrossRef]

28. Sulmon, C.; van Baaren, J.; Cabello-Hurtado, F.; Gouesbet, G.; Hennion, F.; Mony, C.; Renault, D.; Bormans, M.; El Amrani, A.; Wiegand, C.; et al. Abiotic stressors and stress responses: What commonalities appear between species across biological organization levels? *Environ. Pollut.* **2015**, *202*, 66–77. [CrossRef]

29. Pichersky, E.; Gang, D.R. Genetics and biochemistry of secondary metabolites in plants: An evolutionary perspective. *Trends Plant Sci.* **2000**, *5*, 439–445. [CrossRef]

30. Hanada, K.; Sawada, Y.; Kuromori, T.; Klausnitzer, R.; Saito, K.; Toyoda, T.; Shinozaki, K.; Li, W.-H.; Hirai, M.Y. Functional compensation of primary and secondary metabolites by duplicate genes in Arabidopsis thaliana. *Mol. Biol. Evol.* **2011**, *28*, 377–382. [CrossRef]

31. Lehmann, S.; Funck, D.; Szabados, L.; Rentsch, D. Proline metabolism and transport in plant development. *Amino Acids* **2010**, *39*, 949–962. [CrossRef]

32. Di Ferdinando, M.; Brunetti, C.; Agati, G.; Tattini, M. Multiple functions of polyphenols in plants inhabiting unfavorable Mediterranean areas. *Environ. Exp. Bot.* **2014**, *103*, 107–116. [CrossRef]

33. Wagstaff, S.J.; Hennion, F. Evolution and biogeography of *Lyallia* and *Hectorella* (Portulacaceae), geographically isolated sisters from the Southern Hemisphere. *Antarct. Sci.* **2007**, *19*, 417–426. [CrossRef]

34. Bouchereau, A.; Aziz, A.; Larher, F.; Martin-Tanguy, J. Polyamines and environmental challenges: Recent development. *Plant Sci.* **1999**, *140*, 103–125. [CrossRef]

35. Agati, G.; Azzarello, E.; Pollastri, S.; Tattini, M. Flavonoids as antioxidants in plants: Location and functional significance. *Plant Sci.* **2012**, *196*, 67–76. [CrossRef]

36. Hennion, F.; Fiasson, J.L.; Gluchoff-Fiasson, K. Morphological and phytochemical relationships between *Ranunculus* species from Iles Kerguelen. *Biochem. Syst. Ecol.* **1994**, *22*, 533–542. [CrossRef]

37. Hennion, F.; Litrico, I.; Bartish, I.V.; Weigelt, A.; Bouchereau, A.; Prinzing, A. Ecologically diverse and distinct neighbourhoods trigger persistent phenotypic consequences, and amine metabolic profiling detects them. *J. Ecol.* **2016**, *104*, 125–137. [CrossRef]

38. Alcazar, R.; Marco, F.; Cuevas, J.C.; Patron, M.; Ferrando, A.; Carrasco, P.; Tiburcio, A.F.; Altabella, T. Involvement of polyamines in plant response to abiotic stress. *Biotechnol. Lett.* **2006**, *28*, 1867–1876. [CrossRef]

39. Treutter, D. Significance of flavonoids in plant resistance and enhancement of their biosynthesis. *Plant Biol.* **2005**, *7*, 581–591. [CrossRef]

40. Rott, E.; Gross, E.; Schwienbacher, E. Small-scale heterogeneity of *Ranunculus trichophyllus* in Lake Tovel (microhabitat, morphology, phenolic compounds and molecular taxonomy). *Acta Biol.* **2004**, *81*, 359–367.

41. Hummel, I.; Quemmerais, F.; Gouesbet, G.; El Amrani, A.; Frenot, Y.; Hennion, F.; Couée, I. Characterization of environmental stress responses during early development of Pringlea antiscorbutica in the field at Kerguelen. *New Phytol.* **2004**, *162*, 705–715. [CrossRef]

42. Van der Putten, N.; Verbruggen, C.; Ochyra, R.; Verleyen, E.; Frenot, Y. Subantarctic flowering plants: Pre-glacial survivors or post-glacial immigrants? *J. Biogeogr.* **2010**, *37*, 582–592. [CrossRef]

43. Hennion, F.; Walton, D.W.H. Ecology and seed morphology of endemic species from Kerguelen Phytogeographic Zone. *Polar Biol.* **1997**, *18*, 229–235. [CrossRef]

44. Hennion, F. Etude des Caractéristiques Biologiques et Génétiques de la Flore Endémique des îles Kerguelen. Ph.D. Thesis, Muséum National d'Histoire Naturelle, Paris, France, 1992.

45. Lebouvier, M.; Frenot, Y. Conservation and management in the french sub-Antarctic islands and surrounding seas. *Pap. Proc. R. Soc. Tasman.* **2007**, *141*, 23–28. [CrossRef]

46. Lebouvier, M.; Laparie, M.; Hulle, M.; Marais, A.; Cozic, Y.; Lalouette, L.; Vernon, P.; Candresse, T.; Frenot, Y.; Renault, D. The significance of the sub-Antarctic Kerguelen Islands for the assessment of the vulnerability of native communities to climate change, alien insect invasions and plant viruses. *Biol. Invasions* **2011**, *13*, 1195–1208. [CrossRef]

47. Lehnebach, C.A.; Winkworth, R.C.; Becker, M.; Lockhart, P.J.; Hennion, F. Around the pole: Evolution of sub-Antarctic Ranunculus. *J. Biogeogr.* **2017**, *44*, 875–886. [CrossRef]

48. Tiburcio, A.F.; Kaur-Sawhney, R.; Galston, A.W. Polyamine metabolism. In *The Biochemistry of Plants, Volume 16*; Miflin, B.J., Lea, P.J., Eds.; Academic Press: New York, NY, USA, 1990; pp. 283–325.

49. Fujihara, S.; Yoneyama, T. Endogenous levels of polyamines in the organs of cucumber plant (*Cucumis sativus*) and factors affecting leaf polyamine contents. *Physiol. Plant.* **2001**, *113*, 416–423. [CrossRef]

50. Hermant, M.; Prinzing, A.; Vernon, P.; Convey, P.; Hennion, F. Endemic species have highly integrated phenotypes, environmental distributions and phenotype–environment relationships. *J. Biogeogr.* **2013**, *40*, 1583–1594. [CrossRef]

51. Foster, S.A.; Walters, D.R. Polyamine concentrations and arginine decarboxylase activity in wheat exposed to osmotic stress. *Physiol. Plant.* **1991**, *82*, 185–190. [CrossRef]

52. Hennion, F.; Frenot, Y.; Martin-Tanguy, J. High flexibility in growth and polyamine composition of the crucifer *Pringlea antiscorbutica* in relation to environmental conditions. *Physiol. Plant.* **2006**, *127*, 212–224. [CrossRef]

53. Cornelissen, J.H.C.; Lavorel, S.; Garnier, E.; Diaz, S.; Buchmann, N.; Gurvich, D.E.; Reich, P.B.; ter Steege, H.; Morgan, H.D.; van der Heijden, M.G.A.; et al. A handbook of protocols for standardised and easy measurement of plant functional traits worldwide. *Aust. J. Bot.* **2003**, *51*, 335–380. [CrossRef]

54. Smith, M.A.; Davies, P.J. Separation and quantitation of polyamines in plant tissue by high performance liquid chromatography of their dansyl derivatives. *Plant Physiol.* **1985**, *78*, 89–91. [CrossRef]

55. Hayman, A.R.; Gray, D.O.; Evans, S.V. New high performance liquid chromatography system for the separation of biogenic amines as their DNS derivatives. *J. Chromatogr.* **1985**, *325*, 462–466. [CrossRef]

56. Jubault, M.; Hamon, C.; Gravot, A.; Lariagon, C.; Delourme, R.; Bouchereau, A.; Manzanares-Dauleux, M.J. Differential regulation of root arginine catabolism and polyamine metabolism in clubroot-susceptible and partially resistant Arabidopsis genotypes. *Plant Physiol.* **2008**, *146*, 2008–2019. [CrossRef]

57. R Core Team: R: A Language and Environment for Statistical Computing. Available online: https://www.r-project.org/ (accessed on 18 July 2019).

58. Terbraak, C.J.F. Canonical correspondence-analysis—A new eigenvector technique for multivariate direct gradient analysis. *Ecology* **1986**, *67*, 1167–1179. [CrossRef]

59. Legendre, P.; Legendre, L. *Numerical Ecology, Volume 24*, 3rd ed.; Elsevier: Amsterdam, The Netherlands, 2012; p. 1006.

60. Cabin, R.J.; Mitchell, R.J. To Bonferroni or not to Bonferroni: When and how are the questions. *Bull. Ecol. Soc. Am.* **2000**, *81*, 246–248.

MDPI
St. Alban-Anlage 66
4052 Basel
Switzerland
Tel. +41 61 683 77 34
Fax +41 61 302 89 18
www.mdpi.com

Plants Editorial Office
E-mail: plants@mdpi.com
www.mdpi.com/journal/plants